Geographies of Anticolonialism

T0201229

RGS-IBG Book Series

For further information about the series and a full list of published and forthcoming titles please visit www.rgsbookseries.com

Published

Geographies of Anticolonialism: Political Networks Across and Beyond South India, c. 1900–1930
Andrew Davies

Geopolitics and the Event: Rethinking Britain's Iraq War Through Art
Alan Ingram

On Shifting Foundations: State Rescaling, Policy Experimentation And Economic Restructuring In Post-1949 China
Kean Fan Lim

Global Asian City: Migration, Desire and the Politics of Encounter in 21st Century Seoul
Francis L. Collins

Transnational Geographies Of The Heart: Intimate Subjectivities In A Globalizing City
Katie Walsh

Cryptic Concrete: A Subterranean Journey Into Cold War Germany
Ian Klinke

Work-Life Advantage: Sustaining Regional Learning and Innovation
Al James

Pathological Lives: Disease, Space and Biopolitics
Steve Hinchliffe, Nick Bingham, John Allen and Simon Carter

Smoking Geographies: Space, Place and Tobacco
Ross Barnett, Graham Moon, Jamie Pearce, Lee Thompson and Liz Twigg

Rehearsing the State: The Political Practices of the Tibetan Government-in-Exile
Fiona McConnell

Nothing Personal? Geographies of Governing and Activism in the British Asylum System
Nick Gill

Articulations of Capital: Global Production Networks and Regional Transformations
John Pickles and Adrian Smith, with Robert Begg, Milan Buček, Poli Roukova and Rudolf Pástor

Metropolitan Preoccupations: The Spatial Politics of Squatting in Berlin
Alexander Vasudevan

Everyday Peace? Politics, Citizenship and Muslim Lives in India
Philippa Williams

Assembling Export Markets: The Making and Unmaking of Global Food Connections in West Africa
Stefan Ouma

Africa's Information Revolution: Technical Regimes and Production Networks in South Africa and Tanzania
James T. Murphy and Pádraig Carmody

Origination: The Geographies of Brands and Branding
Andy Pike

In the Nature of Landscape: Cultural Geography on the Norfolk Broads
David Matless

Geopolitics and Expertise: Knowledge and Authority in European Diplomacy
Merje Kuus

Everyday Moral Economies: Food, Politics and Scale in Cuba
Marisa Wilson

Material Politics: Disputes Along the Pipeline
Andrew Barry

Fashioning Globalisation: New Zealand Design, Working Women and the Cultural Economy
Maureen Molloy and Wendy Larner

Working Lives - Gender, Migration and Employment in Britain, 1945–2007
Linda McDowell

Dunes: Dynamics, Morphology and Geological History
Andrew Warren

Spatial Politics: Essays for Doreen Massey
Edited by David Featherstone and Joe Painter

The Improvised State: Sovereignty, Performance and Agency in Dayton Bosnia
Alex Jeffrey

Learning the City: Knowledge and Translocal Assemblage
Colin McFarlane

Globalizing Responsibility: The Political Rationalities of Ethical Consumption
Clive Barnett, Paul Cloke, Nick Clarke & Alice Malpass

Domesticating Neo-Liberalism: Spaces of Economic Practice and Social Reproduction in Post-Socialist Cities
Alison Stenning, Adrian Smith, Alena Rochovská and Dariusz Świątek

Swept Up Lives? Re-envisioning the Homeless City
Paul Cloke, Jon May and Sarah Johnsen

Aerial Life: Spaces, Mobilities, Affects
Peter Adey

Millionaire Migrants: Trans-Pacific Life Lines
David Ley

State, Science and the Skies: Governmentalities of the British Atmosphere
Mark Whitehead

Complex Locations: Women's geographical work in the UK 1850–1970
Avril Maddrell

Value Chain Struggles: Institutions and Governance in the Plantation Districts of South India
Jeff Neilson and Bill Pritchard

Queer Visibilities: Space, Identity and Interaction in Cape Town
Andrew Tucker

Arsenic Pollution: A Global Synthesis
Peter Ravenscroft, Hugh Brammer and Keith Richards

Resistance, Space and Political Identities: The Making of Counter-Global Networks
David Featherstone

Mental Health and Social Space: Towards Inclusionary Geographies?
Hester Parr

Climate and Society in Colonial Mexico: A Study in Vulnerability
Georgina H. Endfield

Geochemical Sediments and Landscapes
Edited by David J. Nash and Sue J. McLaren

Driving Spaces: A Cultural-Historical Geography of England's M1 Motorway
Peter Merriman

Badlands of the Republic: Space, Politics and Urban Policy
Mustafa Dikeç

Geomorphology of Upland Peat: Erosion, Form and Landscape Change
Martin Evans and Jeff Warburton

Spaces of Colonialism: Delhi's Urban Governmentalities
Stephen Legg

People/States/Territories
Rhys Jones

Publics and the City
Kurt Iveson

After the Three Italies: Wealth, Inequality and Industrial Change
Mick Dunford and Lidia Greco

Putting Workfare in Place
Peter Sunley, Ron Martin and Corinne Nativel

Domicile and Diaspora
Alison Blunt

Geographies and Moralities
Edited by Roger Lee and David M. Smith

Military Geographies
Rachel Woodward

A New Deal for Transport?
Edited by Iain Docherty and Jon Shaw

Geographies of British Modernity
Edited by David Gilbert, David Matless and Brian Short

Lost Geographies of Power
John Allen

Globalizing South China
Carolyn L. Cartier

Geomorphological Processes and Landscape Change: Britain in the Last 1000 Years
Edited by David L. Higgitt and E. Mark Lee

Geographies of Anticolonialism

Political Networks Across and Beyond South India, c. 1900–1930

Andrew Davies

WILEY

Registered Offices
John Wiley & Sons, Inc., 111 River Street, Hoboken, NJ 07030, USA
John Wiley & Sons Ltd, The Atrium, Southern Gate, Chichester, West Sussex, PO19 8SQ, UK

Editorial Office
9600 Garsington Road, Oxford, OX4 2DQ, UK

For details of our global editorial offices, customer services, and more information about Wiley products visit us at www.wiley.com.

Wiley also publishes its books in a variety of electronic formats and by print-on-demand. Some content that appears in standard print versions of this book may not be available in other formats.

Library of Congress Cataloging-in-Publication data applied for
9781119381549 (hardback); 9781119381556 (paperback)

Cover Design: Wiley
Cover Image: © Jasjit Bajwa/Getty Images

Set in 10/12pt Plantin by SPi Global, Pondicherry, India
Printed and bound in Singapore by Markono Print Media Pte Ltd

10 9 8 7 6 5 4 3 2 1

Contents

Series Editor's Preface

The RGS–IBG Book Series only publishes work of the highest international standing. Its emphasis is on distinctive new developments in human and physical geography, although it is also open to contributions from cognate disciplines whose interests overlap with those of geographers. The Series places strong emphasis on theoretically informed and empirically strong texts. Reflecting the vibrant and diverse theoretical and empirical agendas that characterise the contemporary discipline, contributions are expected to inform, challenge and stimulate the reader. Overall, the RGS–IBG Book Series seeks to promote scholarly publications that leave an intellectual mark and change the way readers think about particular issues, methods or theories.

For details on how to submit a proposal please visit:
www.rgsbookseries.com.

David Featherstone
University of Glasgow, UK
RGS–IBG Book Series Editor

Acknowledgements

Whilst my research has been more or less connected to ideas of anticolonialism throughout my career so far, the research connection to the Pondicherry 'radicals' of the early twentieth century was made possible due to a British Association for South Asian Studies/European Consortium for Asian Field Study/British Academy Fellowship which I received in 2013. As a result, I have run up a huge number of debts to people who have supported this project in innumerable ways. During the Fellowship, I was based at the *Ecole Francais d'Extreme Orient* (EFEO) in Pondicherry, at that time led by Valerie Gillet, and the space and time provided by the EFEO, together with the excellent support I received from Prerana Patel, made the time spent in Pondicherry extremely productive. Much of my time in Pondicherry was spent working at the excellent research library of the *Institut Francais de Pondicherry*, as well as benefitting then and since from the advice and discussions with Kannan M. and Mythri Prasad, as well as from the excellent staff at the IFP Library. In Chennai, the openness and generosity of A.R. Venkatachalapathy, and the ability to stay in the guest house at the *Madras Institute of Development Studies* in early 2015 proved extremely helpful. Elsewhere in Chennai, the staff at the Tamil Nadu State Archives in Egmore, the Roja Muthiah Research Library and the Theosophical Society Library have all provided support at various times during the research. In New Delhi, the staff at the National Archives of India facilitated my research in 2013 during a month long research period.

My colleagues and friends in the Power Space and Cultural Change Research Cluster, and in the wider Department of Geography and Planning have helped to provide a stimulating and happy place to work in the often challenging world of contemporary academia. A sabbatical period in 2015 was essential in providing further time for work on this project collecting data, whilst a further sabbatical at the end of 2018 provided space to allow this manuscript to be completed. Kathy Burrell's support and friendship, especially in our collaboration teaching Third Year Undergraduates on our 'Postcolonial Geographies' module here at Liverpool has been invaluable. Speaking to students on that module, as well as discussing post/de/anticolonialism with our taught and research postgraduate students at Liverpool has certainly helped me develop the ideas present in this book. Elsewhere, Levi Gahman, Kim Peters, Mark Riley, Bethan Evans, Arshad Isakjee, Jen Turner, Lucy Jackson, Pete North, Mark Green, Dani Arribas-Bel, James Lea, Andy Plater, Richard Chiverrell, John Sturzaker, Josh Blamire and Janet Hooke have all listened to me discuss the Pondicherry Gang or the idea of anticolonialism over the last few years with admirable patience, often in times when they'd rather I was talking about something else. Suzanne Yee created the map of South India which appears in the book, and I am grateful to work in a department which still employs her carto-graphic expertise alongside Tinho da Cruz. Elsewhere at Liverpool, Ian Magedera, Nandini Das, Iain Jackson, Deana Heath and Soumyen Bandyopadhyay have all helped with their discussions of South Asia at various points.

A wider geographical and South Asian circle of scholarship has also proved of huge benefit. I have presented aspects of this book in seminars at UCL, Glasgow, Leicester, Aberystwyth, Sheffield, Jawaharlal Nehru University, New Delhi, and at various sessions of the RGS–IBG Annual Conference, as well as at conferences in Manchester and in public lectures in Liverpool. All of the comments received have proved extremely helpful in honing my arguments. More specifically, Stephen Legg, Tariq Jazeel, Gavin Brown, Federico Ferretti, Ole Laursen, Peter Hopkins, Rhys Jones, Jenny Pickerill, Santana Khanikar and many more have provided support for this project and my ideas more generally. Thank you all.

Archival staff at the Centre of South Asian Studies, Cambridge, the British Library, London and the National Archives in Kew have all helped with documentation. Manu Goswami helpfully provided a copy of a hard to find article, whilst Jessica Nammakkal kindly agreed to let me reference her twitter discussion about Auroville.

My journey into geographical anticolonial thought would not have occurred without being supervised on a PhD by Richard Phillips and David Featherstone. Hopefully this manuscript is evidence that all of those cups of proper tea added up to something other than theft. Both have also been exemplary mentors as my career has developed, and Dave has been a thoughtful and generous editor for this book. I would also like to extend wholehearted thanks to the two anonymous reviewers of this manuscript, whose advice and commentary helped to sharpen the work as a whole. At Wiley, Jacqueline Scott has been a model of patience and knowledge. Elisha Benjamin and Navami Rajunath have been excellent production editors.

Friends in Liverpool have likewise had to learn a lot about anticolonial radi-calism in India and elsewhere over the past few years, often when they didn't really want to. Thanks to them for their patience – particularly Frazer, Nicki, Steven, Steve, Amy, Kev, Paul, Nick, Lorraine and Paul as well as others too numerous to name. Mam, Dad and Nathan have also been a huge support across these years. Lastly, but most significantly, Caroline has had to deal with my anti-colonial obsession on a daily basis for many years, and has provided much needed emotional support, not least helpfully wiring cash to me when I realised my debit card had expired five days into a four-week trip to Pondicherry in 2015. I promise never to write a book while we're moving house, to help more with the redecorat-ing, and to spend less time with my Indian archival 'family' in the future.

Author's Note

Pronunciation and Transliteration

Tamil, as a Dravidian language, is distinct from the Sanskrit-derived languages of Northern India. I have attempted to use the current standard transliterations throughout this book, but there are areas of inconsistency. For instance, V.V.S. Iyer is often also spelled as V.V.S. Aiyar. I have opted for the latter spelling in this book, but there is no formally recognised preference here as far as I am aware. It is also worth noting that 'zh' is a distinct letter within Tamil and other Dravidian languages. When pronounced, it sounds something like an English 'L', but with a retroflex position of the tongue – produced by holding your tongue close to, but not touching, the roof of the mouth, and trying to pronounce an English 'z', while rolling the tongue slightly. Strictly, the word 'Tamil' should be spelt Tamizh in English, although I have avoided this to minimise confusion, but elsewhere, have used zh (for example, in Dravida Munnetra Kazhagam, one of the main Tamil political parties of the last 50 years).

Glossary and List of Abbreviations

Bharat/Bharata	Alternative name for the geographical region of India, derived from Sanskrit and the ancient texts of the puranas
CID	Madras Presidency Criminal Investigation Department. The regional intelligence gathering service for Madras, which reported to the Government of Madras, which then reported to the GoI
DCI	Department of Criminal Intelligence, Government of India. The office in charge of gathering intelligence for the Government of India, not to be confused with the CID above

DMK	Dravida Munnetra Kazhagam (Dravidian Progress Foundation), one of the major postcolonial Tamil political parties. Its main opponent, the AIADMK (All India Anna Dravida Munnetra Kazhagam) was founded when the party split in the early 1970s
GoI	Government of India
Hind	Alternative term for the geographical territory of India. Derived from Persian, but also Greek writings on India, as the lands around the Indus river, but which became synonymous with India as a whole over time
INC	Indian National Congress
IPI	Indian Political Intelligence
LAI	League Against Imperialism
NAI	National Archives of India, New Delhi
Satyagraha	Lit. Truth-force, or firm adherence to the truth
Swaraj	Lit. Self-rule
TNA	Tamil Nadu State Archives, Chennai

Chapter One
Post? Anti? De? Why Anticolonialism Still Matters

Introduction

From around 1906 to 1915, a small group of Indian men (with an even smaller group of women) found themselves in the small French territorial enclave of Pondicherry on India's southeastern coastline. The men, the majority of whom were from Tamil-speaking regions of India, were the core of an emergent anticolonialism in Southern India. Already, they had been involved in a number of revolutionary nationalist activities, and all were identified as being parts of an 'extremist' faction amongst the Indian National Congress (INC), the main legal organisation by which Indians could agitate for change within British-governed India. By moving to Pondicherry, the group escaped from immediate prosecution or imprisonment, but moved into a space of exile, shut off and contained within the small confines of the city and with limited means to earn money and survive.

Despite these restrictions, Pondicherry became an important nodal point in an international network of anticolonial agitation which stretched to London, Paris, New York, San Francisco, Tashkent, Constantinople, Berlin and Tangiers to name only a few. Revolutionaries passed through Pondicherry, were not only trained there in the techniques of anticolonial revolution, but also smuggled guns and revolutionary materials into the rest of India through the city's port. However, being confined to Pondicherry, as well as the changing political situation in India and globally by the outbreak of World War One, meant that the group was never able to agitate effectively and to create a mass movement in South India, and by

Geographies of Anticolonialism: Political Networks Across and Beyond South India, c. 1900–1930, First Edition. Andrew Davies.
© 2020 Royal Geographical Society (with the Institute of British Geographers). Published 2020 by John Wiley & Sons Ltd.

1915, Pondicherry's moment as a centre for political radicalism had gone. Despite this, for such a small group operating in often distressed circumstances, the radicals who found themselves in Pondicherry played an important role in shaping the wider geographies of anticolonialism which were emerging globally in the early twentieth century.

It is this group, who came to be known variously as the Pondicherry 'Gang', 'group', 'anarchists' and a variety of other names by the British authorities who spied on them whilst they lived in Pondicherry, who form the focus of this book. What the book argues is that although a relatively minor moment in the struggle for India's independence, the activities of the Pondicherry 'Gang' are of vital importance in understanding how anticolonialism was a diverse set of activities, which crucially were productive of new geographies of the world. The Pondicherry group of anticolonialists was never really as cohesive a group as the colonial authorities made them out to be. However, the individuals who moved through Pondicherry illuminate how anticolonialism was dynamic and heterogeneous in its attempts to resist imperial and colonial domination.

This emphasis on anticolonialism is particularly important as the legacies of colonialism, and how we deal with them is increasingly visible in academic and public life in the twenty-first century. Social and political movements have emerged over the past decade to challenge orthodoxies and colonial assumptions which, alongside long-existing struggles, have shifted some of the face of public discussion. In North America, the indigenous struggle over the Dakota Access Pipeline at Standing Rock brought together indigenous communities and environmental activists in ways which reworked existing solidarities, and the fight for racial justice, with its often underplayed links (in mainstream commentaries at least) to colonial racial hierarchies, have meant that the anti-confederate statue movement, Black Lives Matter and the NFL Anthem/Colin Kaepernick protests have all worked to destabilise many hegemonic structures of power and domination. In South Africa, the Rhodes Must Fall movement at Cape Town University expanded to a nationwide and eventually transnational struggle to recognise the colonial legacies that are embedded within higher education institutions across the world (Bhambra, Gebrial and Nisanclogu 2018; Rhodes Must Fall Oxford 2018). Elsewhere in Europe, activists and intellectuals have been developing a range of responses, from policy documents to museum galleries to policy reports/books, to highlight the complicity of nations from Scandinavia to Belgium to Germany in colonial projects which are often downplayed compared to British or French colonialism.

This process has become so clearly important to academia that the need to be 'decolonial' or to 'decolonise' our education systems and wider societies is frequently used as a rallying cry for a variety of social justice movements. In terms of geography, it is notable that the 2017 Royal Geographical Society Conference theme was 'Decolonising Geographical Knowledges'. At the same time, there has been a predictable backlash against these movements. From the broadly 'right-wing' sections of society, university students and lecturers, often when they are women, or people of colour, or most often both, have been attacked for, amongst other things, being 'snowflakes' who are too sensitive about historical injustices and need to toughen up to the realities of a harsh world, or for attacking and seeking to remove traditional (read: 'white') authors and intellectuals from disciplines. These attacks often wilfully ignore the contexts and nuance of the arguments for creating a more inclusive and diverse education system (and indeed wider world).

This book is written then at something of a 'decolonial' moment where the varied practices, experiences and legacies of colonialism are at the forefront of many debates in society. Why then, is this book titled 'Geographies of *Anti*colonialism'? As will become clear throughout this introduction and the rest of the book, whilst the decolonial and the postcolonial are undoubtedly vital and important to the ongoing struggle against colonial forms of domination, my argument is that geography would be well served not to 'forget' the anti-colonial as an equally important, yet subtly different register for understanding the practices and experiences of resistance to colonialism/imperialism. As Mary Gilmartin, in the discussion surrounding her lecture for the journal *Political Geography* at the RGS–IBG conference in 2017 argued, we should be mindful of what the struggles against colonialism in the past and present actually offer us, and what anticolonialism's rigorous commitment to dismantling the political structures of colonialism gives us alongside the equally important cultural, epis-temological and ontological strategies which form a core aspect of postcolonial and decolonial approaches (see also Gilmartin 2009; Naylor et al. 2018 for more by Gilmartin on post-/de/anticolonialism and geography). For the rest of this introductory chapter, I will set out the ways in which geographers have engaged with these terms. In particular, through this chapter, and on into the next chapter which deals with anticolonial thought in more detail, I want to argue that, whilst all three terms, and their related schools of thought are 'against' and therefore 'anti-colonial', geographers and academics are, as Gilmartin suggested, missing out on something vitally important if we ignore or downplay the anticolonial in these more contemporary movements. In the suitably imperial/colonial sur-roundings of the Ondaatje Theatre of the RGS building in South Kensington, London, I remember having to suppress a cry of 'Yes! Someone else gets it!' when Mary said this.

Crucially, my point is that anticolonialism is not just an aspect of the past, or a historical relic, something which it is often relegated to – for example, in Robert J.C. Young's recent (2015) primer on the variety of approaches to colonialism, the chapter on anticolonialism deals exclusively with movements that struggled against formal practices of colonialism in the past – and therefore, anticolonialism is something that is discussed in the past tense, rather than something which is ongoing. This book is my attempt to think through what that response could be, and to (hopefully) open up ground for other attempts to understand the myriad ways in which anticolonial geographies were and are relationally produced in practises which fruitfully intersect with post- and decolonial approaches to geography. It is then, something of a beginning and hopefully opens up some space for similar discussions in the future about different anticolonial geographies.

Before moving on to discuss some of these terms and their evolution in geographical thought in more detail, it is worth mentioning a small detail about terminology. For a long time in postcolonial theory, and in postcolonial geography, there were a lot of debates about whether there should or should not be a hyphen inserted in the term (i.e. post-colonial). Whilst these debates have largely disappeared, it is still important to clarify my particular use of terms here. I tend to use postcolonial as I think that the hyphen tends to create an emphasis on the experience of colonialism as something that is 'past' or over. Removing the hyphen, to me, not only allows a sense of the continued existence and legacies of colonialism to be maintained but also to indicate that the term is continuing to work beyond and against colonialist practices. Likewise, I do not use the term 'anti-colonialism', rather choosing to use anticolonialism. Whilst my objection to the postcolonial lies largely, but not wholly, in its creation of a certain type of temporality, I use anticolonial to break down the sense that anticolonialism is only a negative reaction to the colonial situation. Removing the hyphen, to me, not only helps to both show that practices of anticolonialism were always situated and calculated responses to the specific colonial encounters that produced them but also gives these practices the space to exceed those circumstances. As will become clear throughout this book, the development of a number of responses to colonial and imperial rule involved not only negating and contesting the dominant narratives of colonialism but also showing how to imagine a future world beyond them that would be more than, to use the terms of John Holloway (2002), a shout 'against' the colonial system. With this clarification in mind, the rest of this introduction sets out the evolution of geography's varied engagements with resistance to the colonial and the imperial.

'Postcolonial' Geographies?

The impact of postcolonialism on geography is in danger of being overlooked, given the present desire to embrace decolonial approaches. The supposedly passé or elitist nature of postcolonial theory has meant that the term has come under a number of vituperative attacks in recent years – see, for example Hamid Dabashi's (2015) short but damning criticism of Homi K. Bhabha's 'useless bourgeois post-modernism' (p. 7). However, despite the tensions inherent in the term, it is important to state that without the development of a variety of postcolonial approaches to geography since the early 1990s, the discipline as a whole would look remarkably different. To many postcolonialists, it was the publication of Edward Said's (2003) *Orientalism* that mark the emergence of a distinct postcolonial theory. It is, of course, problematic to assign a specific moment to the emergence of such a diverse range of thought, and, as Marxist critics of postcolonialism have noted, Said's orientalism only signalled the emergence of an increasingly post-modern trend within postcolonial studies (Kaiwar 2015). However, for geography, postcolonialism emerged as an approach later on, around the same time as the 'cultural turn' in Geography. Postcolonial geographies are fundamentally an attempt to understand the intersections between colonialism and space. As Blunt and McEwan argue in the Introduction to *Postcolonial Geographies* (McEwan and Blunt 2002, p. 1), 'postcolonialism and geography are intimately linked. Their intersections provide many challenging opportunities to explore the spatiality of colonial discourse, the spatial politics of representation, and the material effects of colonialism in different places'. Interestingly, Blunt and McEwan argue that many of the chapters in their edited collection were working towards decolonising knowledges and the articulation of colonial discourses in the past and the present. The significant overlaps between post-/de/anticolonial approaches have therefore always been present. McEwan and Blunt's book, as well as Sidaway's survey chapter in the same collection and the chapter on postcolonial geography in Blunt and Wills (2000), marked an important point in the official recognition of postcolonial approaches to geography. Prior to this point, whilst there were a number of geographers who were writing about postcolonial issues (Blunt and Rose 1994; Phillips 1997; Radcliffe 1997), it was in the early 2000s, that postcolonialism was placed firmly as a topic of concern to all geographers. This shift was something that happened rapidly in my own personal experience, where, as someone who graduated in 2002 from an undergraduate degree which had no explicitly postcolonial content during the entire programme, I suddenly encountered an invigorated and vibrant postcolonial approach to geography when I returned to the academy to undertake an MA in 2004/2005.

However, for many geographers around this time, postcolonial approaches utilised the various theoretical 'tools' provided by the scholars who were working through the details of postcolonial thought in cultural studies and elsewhere. It is therefore not surprising to see the well-established work of cultural and literary postcolonial approaches being used as key touchstones by geographers at this time, and the likes of Ashcroft et al. (2000) and Loomba (1998), as well as key thinkers such as Edward Said or Homi K. Bhabha or Gayatri Spivak, were often key citations in this work. It is also from criticisms of the supposedly 'poststructural' and culturally inflected nature of these authors that the initial critiques of postcolonialism and postcolonial geographies emerged, particularly from scholars such as Arif Dirlik and Aijaz Ahmad. Whilst these critiques are well known, it is worth examining them here as they set out the frameworks for much of the critique that was to follow, and which has coalesced around the drive for 'decolonial' approaches to geography today, but with some important differences. Both Ahmad and Dirlik are broadly Marxist in orientation, and are also deeply sceptical of the post-modern or post-structural nature of postcolonial thought in the 1990s. For Ahmad (1995), this meant that term had evacuated much of the initial idea of the postcolonial, which had emerged out of the anticolonial nationalist struggles of the mid-twentieth century – where the postcolonial was closely linked to temporality and was a marker of the political project to create a nation, or even a world, after colonialism. To Ahmad (1992), the emergence of postcolonial literary theory had moved this struggle away from the political economy of imperialism/colonialism, into the realm of the cultural 'superstructure' of Marxism. This intellectual shift meant that, on the one hand, postcolonial theory became increasingly bourgeois and detached from political struggle, but on the other, was also prone to making and repeating assumptions about the postcolonial world which were rooted in colonialist framings. In particular, Ahmad was famously critical of the range and scope of the literatures that many postcolonial literary theorists were working with, most notably in his polemic on Frederic Jameson's categorisation of 'Third World literature' and the colonialist assumptions which underpinned that conception.

Following a similar line of argument, according to Arif Dirlik (1999), the postcolonial had become by the late 1990s so diffuse a term as to be almost worthless. By moving away from its original domain of resistance to actually existing colonialism into a more nebulous area where ethnicity and representational politics seemingly took precedence over gendered and classed politics, postcolonial theory, it was argued, not only missed the terms of the game, but failed to achieve much in terms of concrete change for those who suffered the negative effects of colonial rule, and often served to reinforce existing colonial relations:

> Conceived to combat ethnocentrism and racism, postcolonial discourse ironically contributes presently to the racialisation and ethnicisation of the languages of both critical intellectual work and politics – with liberal intentions, no doubt, but at the risk on the one hand of covering up proliferating problems of social inequality and oppression whose origins lie elsewhere, and, on the other hand, of contributing to the consolidation of the very ethnic, national and racial boundaries that it is intended to render porous and traversible. (Dirlik 1999, p. 153)

By its expansion beyond its immediate terms of reference as a temporal and a political issue, postcolonialism 'retreated' from two of its earlier key concerns – 'nation' and 'class' (Dirlik 1999, p. 150) and became too concerned with cultural and representational issues. Crucially to Dirlik, these cultural issues were not missing from the earlier period, but instead formed part of a wider political movement to achieve national liberation. The debt to Marxism in these criticisms is clear, but they do highlight the perceived limits of postcolonial critique as post-modern and focussed on the representational at the expense of the formally 'political'.

By the mid-2000s, postcolonial approaches to geography had become established across the discipline to such an extent that Joanne Sharp's textbook

Geographies of Postcolonialism (Sharp 2009) was published, marking for the first time a dedicated text for undergraduate students on the core aims and scope of postcolonial geographies. Within geography, especially after the publications noted earlier, there continued to be debates about the utility and use of postcolonial thought in various sub-disciplines. Despite geographers' awareness of the criticisms of postcolonialism's intellectual diffusion noted earlier, as Sidaway (2002, p. 12) argues, the opening up of postcolonial thought allowed the creation of an 'open constellation of meanings associated with the term'. Indeed, postcolonial geography has proved a ripe ground for challenging a number of intellectual orthodoxies. In political geography, David Slater (2004) took traditional notions of geopolitics and used postcolonial ideas to interrogate their assumptions. In an intervention into geopolitics but from a feminist subaltern perspective, Joanne Sharp (2011, 2013) has argued for alternative centres of geographical knowledge production such as Nyerere's Tanzania to be recognised. Elsewhere, Jennifer Robinson (2003, 2006, 2016) applied postcolonial ideas to urban studies, challenging the Eurocentric basis of many urban models, and this has opened up a whole set of debates about how to imagine a plural urbanism for the twenty-first century, but which have also had important consequences for how we imagine space/geographies more broadly. Likewise, development geographies have been clearly influenced by the postcolonial imperative (Radcliffe 2005; Power, Mohan and Mercer 2006).

Although not true of all of the aforementioned examples, much of the work in postcolonial geographies was avowedly interested in the representational. Given Said's inherently geographical understanding of the production of the orient and its 'imagined geographies', there was undoubtedly much to be done in exposing and challenging these geographies, not least because of institutional geography's complicity in imperial and colonial projects which reproduced these ideas. Whilst earlier work by the likes of Blaut (1993) had exposed the epistemological coloniality at the heart of geographical thought, the emergence of postcolonialism allowed a generation of geographers to show how coloniality was produced and consumed by the subjects of imperialism/colonialism, from advertising through to expos and fairs. However, by the mid-2000s, the emergence of non-representational, or more-than-representational, theories, had begun to alter postcolonial approaches. As a result, there has been a wave of work which has begun to think in post-human terms about the postcolonial, enrolling non-human subjects into our understandings of the world. This has been driven at least in part by Spivak's (2003) discussion of planetarity, which has had important consequences for how geographers have used the postcolonial to understand ontology as well as epistemology (Jackson 2014).

Postcolonial geography has remained one of the discipline's most vibrant areas in recent years, bringing together and challenging many problematic categorisations and rethinking terms which have seemingly been forgotten about or become passé in other disciplines (for example the work done on the subaltern in geography in recent years (Featherstone 2015; Jazeel and Legg 2019)). There is then a rich and continuing debate which is alive in the discipline about understanding how links to colonialism are still present. Postcolonial geographies have therefore been crucial to broadening the scope and challenging the Eurocentricity of geographical knowledge production. This project is still demonstrably ongoing, yet the Marxist-inspired criticisms of the supposedly bourgeois, or at least overly theoretical and post-structural nature of postcolonial theory have continued to hang over the sub-discipline of postcolonial studies more broadly, and in geography, this is no different. Whilst I do not necessarily agree with many of these critiques, they have been at least partially responsible for the shift towards calls to 'decolonise' our societies rather than to think postcolonially about them, and so the next section of this introduction moves on to these approaches which are now prominent beyond academia.

Decolonial Geographies?

Postcolonial thought's resistance to conceptual discipline has meant that it is unable to attach a specific set of political values to itself. Whilst it is in general committed to an ethics of liberation and against oppression, the wide range of different colonial/imperial situations means that many postcolonial theorists are unwilling to corral the ethics of the postcolonial into one particular frame. This has meant that postcolonialism remains resolutely committed to the theoretical. This may be a commitment to 'minor' theory/ies (and more on this in the next chapter), but unlike Marxism or feminism, postcolonialism does not have a distinct praxis *per se*. This is undoubtedly one of the reasons why the term was targeted so heatedly by Marxist critiques in the 1990s. Despite the openness of the term, and the ability of intersectional politics, such as Sharp's feminist postcolonial geopolitics which was mentioned earlier, to be fruitful avenues of academic endeavour, the desire for an openly practical resistance to colonialism, and particularly one that can be used in the present, has driven the emergence of an increasingly visible movement to 'decolonise' our world. This move has driven a debate that has not necessarily split postcolonial geographers, but has at least provoked a new wave of thought and engagement with the terms by which we make sense of the colonial world around us.

Decolonial approaches, whether in geography or more broadly, most often trace their lineages back to anticolonial struggles rather than the literary work of Edward Said et al. Most often, Frantz Fanon (1963, 1964) is invoked as an important figurehead in this process, but equally, the works of Ho Chi Minh, Amilcar Cabral, Sylvia Wynter, and others could be seen as originary. As it is broadly understood now, there are two overlapping but related approaches that gave rise to decolonial approaches (see also Bhambra 2014). The first strand is the emergence of a distinct Latin American engagement with the postcolonial theory which shifted debates that had, in anglophone and literary postcolonial debates, been shaped largely by discussions between European and South Asian scholars. This is most often framed through discussions of the 'Latin American Subaltern Studies Collective' (LASS), a term used in contradistinction to the Subaltern Studies collective which coalesced around studies of South Asian politics in the late 1970s to early 1990s (and more on these in Chapter Three). This term is not unproblematic as it suggests that the debates in South Asia and Europe were precursive and potentially originary of this Latin American intellectual movement (Mignolo 2000). In fact, the LASS agenda was related to but distinct from the Subaltern Studies Collective's project, not least because it emerged in relation to the specific colonial milieu of Latin America. Without wishing to reduce a huge range of writing to a few key figures, it is often clear that writers such as Anibal Quijano, Walter Mignolo and Maria Lugones have provided important practical and conceptual tools for academics. For the likes of Quijano and Mignolo, debate circulates around how the present is and remains colonial. Quijano's pairing of 'coloniality/modernity' (Quijano 2007) as a single term, emphasises that we cannot understand modernity without understanding how it continues to be shaped by colonial process, and thus the condition of our times is its very 'coloniality'. Mignolo's 'colonial matrix of power' (2011) emphasises a similar point, but discusses how dominant, colonial framings and codings of knowledge and subjectivity; race, gender and sexuality; economic management, and; systems of authority; all work together to maintain coloniality in the present world system. Lugones (2010) takes an explicitly feminist approach to this work, showing how patriarchy is embedded within coloniality, with the colonial, hetero-sexual male becoming the 'perfect' marker of civilisation, however, this hierarchy is also complicated by the racialisation and heteronormativity of women's bodies, where those who were deemed aberrant are pushed further down the hierarchy based on their intersectional position. Lugones is clear that the task of this deco-lonial feminist approach is rooted in praxis – it is a 'lived transformation of the social' (Lugones 2010, p. 746). This decolonial approach makes it a core concern that although formal imperialism/colonialism may be over in much of the world, coloniality is not, and challenging the categorisations of modernity/coloniality remains a crucial challenge.

The second strand of decolonial thought has emerged from the very present, lived, experience of coloniality amongst indigenous communities, particularly, but not exclusively, those from North America. Notable here are attempts to challenge the epistemological and methodological foundations of research and academic prac-tice more generally (Smith 1999) through to Fanonian-inspired discussions of the racial politics of being identified as 'indigenous' (Coulthard 2014). Again, the core aim and discussion here is how the colonial is still experienced and confronted on a regular basis. Within geography, this has spurred a range of developments, from work on settler colonial masculinities (Gahman 2016), post-foundational accounts of indigenous communities' engagements with the legal system (Sparke 2005) and understanding occupation as a key strategy of indigenous communities so that it prefigures the supposed territory-claiming novelty of the Occupy! Movement (Barker 2012). This second approach is important because it makes clear that the struggle for decolonisation is not only an epistemic one, where the colonial foundations of knowledge production need to be dismantled, but is also fundamentally about struggles over land and territory in the colonial present. The very real and immediate need to consider how colonialism is made manifest in the present, and how to resist it in the specific contexts in which it arises (Tuck and Yang 2012).

This situates decolonisation/decoloniality as very much a concern for the world we inhabit today. Unlike postcolonialism's uneasy relationship with the past or the use of the hyphen, decolonial theory and practise marks an attempt to challenge and deconstruct the practises of imperialism and colonialism that still exist today. This has meant that decolonisation has become a key marker of many social and political movements in the twenty-first century. For much of the last 50 years, indigenous and anticolonial movements across the world have long used the language of colonialism, the hegemonic discourses of the time meant that such claims were often seen as hyperbole – witness the shift in the 1980s in exiled Tibetan political discourse away from the language of colonialism and independence from China towards a seemingly more 'pragmatic' tone of autonomy within China in the hope of gaining some traction in the realpolitik of the Tibet Issue. However, movements like Occupy!, the aforementioned indigenous activism, the Ejército Zapatista de Liberación Nacional (EZLN), and Rhodes Must Fall (amongst many others) mean that decolonial campaigns have become news in recent years. To be sure, this has often, especially in the United Kingdom and the USA, been at the expense of the campaigners themselves – see, for example the various campaigns to decolonise UK universities, where students and academics, especially those who are non-white and/or female or non-cis, and have often been targeted by sections of the press for attacks for their supposedly over-sensitive behaviour. This is a mark of the various 'culture wars' that have been taking place so far in the twenty-first century. These have long histories in themselves, but the crucial point here is that decolonisation has provided a set of tools by which people are actively seeking to challenge the ontological and epistemological frameworks which many take for granted and which continue to produce conditions of coloniality for many in the world.

This emergence has certainly intersected with much of the writing in postcolonial geographies, especially since the late 2000s. Discussions of planetarity, although owing a debt to Spivak, also clearly intersect with decolonial ideas, and the ontological turn within geography has also provided some fertile ground for decolonial and postcolonial intersections (Jackson 2014). This has also meant that we continue to have to interrogate the practices of geographers as professional researchers. Whilst there have been numerous studies of the troubling past of many geographers during the colonial era (Kearns 2010), there are also still concerns about much more recent practices of colonialism taking place within geography (Wainwright 2012). It is from these concerns that a debate around the 2017 RGS–IBG conference theme of Decolonising Geographical Knowledges took place. In particular, members of the Race, Culture and Equality (RACE) Working Group wrote a number of pieces about the limits of Geography to actually decolonise itself (Esson et al. 2017). These, along with a number of sessions at the RGS–IBG, opened up the terms of discussion for geographical knowledge production in the present and future. Crucially, Esson et al. pointed out the tension of having a conference theme within the spaces of the RGS–IBG, an institution which is indelibly marked by its colonial past. The danger of having the conference theme based on decolonising geographical knowledge was, as Esson et al. rightly point out, that such a theme would not lead to a sustained and rigorous interrogation of the colonial structures at work within the RGS–IBG itself (as well as in academia more generally), but would rather allow the RGS in future to say it had worked to confront these issues, and that would somehow be enough.

The issues raised by decolonisation then are at the forefront of many academic and societal debates at present. The need to address the concerns of colonised people and to rigorously confront the colonial matrix of power within geography (and beyond) is vital to the discipline. However, the imperative of decolonisation means that it is constantly driving towards thinking about how to decolonise in the present to work towards a future where it may be possible to work outside coloniality. This present-centredness is undoubtedly vital, as the debates around the RGS in 2017 and since have shown. However, as I noted earlier, there is a danger that we turn away from past *anti*colonialisms or forget about the lessons that we may learn from them, and this is what this book seeks to address.

Geographies of Anticolonialism?

The overviews above set out how post- and decolonial approaches have had significant impacts upon geographical knowledge production over the past three decades. Why then does this book use 'anticolonialism' as its key term? As Stephen Legg has recently noted, 'colonialism' could be defined as 'practises within colonies' (Legg 2017, p. 347). However, this could be read as situating colonial practises as only taking place within territories that are colonised – and not in places like the colonial centre of power in the metropole – the Colonial Office in London, for example. Likewise, one of the key aims of this book is to think about how the spaces of anticolonialism were not limited to only 'colonies'. We will see how a series of activities in South India spurred a number of events that stretched across the world. Therefore, anticolonialism is a broad range of activity that can be seen to be acting against the similarly broad range of practices and by-products of colonial rule that include racism, militarism, resource exploitation, land dispossession, and so on. This can include resistance to internal and external colonialisms (i.e. within/outside a nation-state's domestic borders, respectively), which also encompass a number of different practices.

Anticolonial geographies are therefore closely linked to, but different from post- and decolonial approaches. They differ from postcolonial approaches because of their explicitly political nature. It is also a somewhat broader concept than decoloniality, which is often (and necessarily) rooted/routed through the specific circumstances of colonialism in particular places/spaces – something like a broad stance rather than a toolkit for precise circumstances. The anticolonial then can act as something of a bridge between the postcolonial and the decolonial, not only suturing them together but also highlighting different practices and activities which are allied to these approaches, but may be ignored or occluded by them. As a result, I argue through the rest of this book that taking an anticolonial approach remains something that will offer a different perspective to geographers. Crucially, this is not an approach that is only historical. Thinking anticolonially about space and place remains relevant to understanding the present and should not be something that is relegated to the past simply because much of the world has now been formally granted political independence from colonial rule.

In the next chapter, I go into more detail about what exactly 'anticolonialism' as an approach to thinking, and importantly thinking geographically, could mean. I not only think through some of the key writings on and by anticolonialists but also develop a sense of what the geographies of anticolonialism could look like. The book does this by thinking through a series of geographies related to the activities of the so-called 'Pondicherry Gang'. As noted earlier, the 'Gang' as such never really existed as a coherent group of people, apart from one short period of activity between 1910 and 1911 – instead, Pondicherry acted as a hub for a number of anticolonially minded individuals to pass through, and even then, for only a short period of time. I re-utilise the term Pondicherry 'Gang' to emphasise the connections between the men (and women) who passed through the spaces of anticolonialism connected to Pondicherry which also reworks the denigration inherent in the colonial use of the term to instead suggest the solidarity and friendships which were shared by the group. This arguably creates too much of a sense of connection between the often disparate groups of people connected to Pondicherry. However, it also stresses the fact that these individuals did share connections that bound them together in their opposition to colonial rule, and which had dramatic consequences for most of their lives.

The historical context of these activities is also important because, as I will show, the various anticolonialisms which took place in and surrounding the Pondicherry 'Gang' show how diverse anticolonialism was in the past. This has important effects in terms of understanding things like the evolution of politics and nationalism in India today, but does more than this in that it also shows how different political subjectivities such as anarchism or spiritual reformism were a part of these movements. When today decolonial and feminist approaches are rightly asking us to think intersectionally about our present, we can look to historical examples such as the 'Gang' to see how these anticolonialisms played out in the past. This is important especially as the activities of the 'Gang' were often destined to end in failure. If we are to decolonise our world, an ability to learn from and think through the challenges and mistakes of the past attempts to decolonise is key to this process.

To establish how the book will do this, it is necessary to say a little here about the structure of the book to come. As mentioned, I set out the anticolonial approach to geography which is central to my argument in the next chapter (Chapter Two), which involves a discussion of some of the key theoretical framings of anticolonial thought. In Chapter Three, I think through some of the core approaches to understanding the historical geography of the struggle for Indian independence which are most relevant to this book. These include discussing ideas about nationalism and internationalism, non-violence and revolutionary violence, and the subaltern. These are important as they continue to develop some of the themes of anticolonialism developed in Chapter Two, but show how anticolonialism was articulated within these specific places. However, as well as continuing the engagement with more theoretical concerns, later in the chapter I begin to introduce some of the history of Southern India which is necessary to place the later chapters in their context. This is important not only in terms of 'setting the geographical scene' and providing some regional context to the later chapters. Instead, an important consideration for this book is to provide a space by which the political activities which took place in Tamil-speaking areas of South India are properly recognised. Traditionally, South India has been characterised as much less militant during the freedom struggle than areas like the Punjab or Bengal. This book shows that this narrative is misplaced, and in fact, South India has its own distinctive place within the history of the freedom struggle.

After these opening chapters, the next four chapters form the historical heart of the book, and each addresses a specific space of anticolonialism though one individual who was either loosely affiliated to the Pondicherry 'Gang' or was active in spaces associated with it. The chapters themselves proceed relatively chronologically, and Chapter Four examines the emergence of *swadeshi* forms of radicalism in South India. *Swadeshi* activism was a complex arrangement of different political strategies which sought to encourage the economic development of India and the reduction of its reliance on foreign-produced (predominantly British) goods. It marked the first nationwide wave of radical protests against British rule in the early twentieth century but also had a number of regional variations. Specifically, this chapter explores the maritime-affiliated spaces of anticolonialism which were created when V.O.C. Pillai attempted to challenge British maritime supremacy of the Indian Ocean by setting up the *Swadeshi* Steam Navigation Company in Tinnevelly district in the far south of India in 1906. The chapter investigates how Indian 'nationalism' was not terra-centric, instead expanding to wider imaginaries about both the industrial development required not only for a future independent Indian nation but also about how this nation would require maritime dimensions. Crucially, the symbolic power of maritime steamship enterprises played into shaping South India's largest-scale industrial nationalist protests, and which marked a key moment in the emergence of anticolonialism in the region.

Chapter Five continues the development of the place-based geographies of Southern India through a discussion of the Tamil poet and polemicist Subramania Bharati. Bharati's life course connects the emergent anticolonial nationalisms caused by the *Swadeshi* Steam Navigation Company to wider currents of political activism in Madras Presidency, the largest British-governed territorial jurisdiction in South India. Bharati was the first political exile to move to Pondicherry, escaping from Madras under the threat of arrest due to his 'seditious' writing and publishing. However, Bharati was a key cultural figure in Tamil-language circles. Today, he is largely credited with reinvigorating Tamil literature, creating a new style of writing that was suitable for political commentary, and establishing a number of standards which continue to shape Tamil political culture. This not only re-emphasises the intersections between space, culture and the political as they played out in South India but also tells some more complex stories about how the spatial politics of anticolonialism are actually mobilised and articulated.

Chapter Six challenges some of our ideas about what counts as 'political' by thinking about the intersections between the political and the spiritual by looking at the life of 'Sri' Aurobindo Ghose, an 'extremist' member of the INC who, when he moved to Pondicherry in 1910 supposedly 'gave up' politics for a life of spiritual retreat into his own practice of integral yoga. This brings into conversation decolonial ideas about the framings of coloniality/modernity, specifically challenging the idea that Aurobindo's move into yogic asceticism was somehow 'non-political' – which tends to be the dominant framing which has been applied to him ever since this move baffled the colonial authorities – at the time, Aurobindo was the most wanted man in India, and his turn away from 'the political' caused a degree of confusion about his motives. However, I argue that Aurobindo's move here is less a shift away from politics, but rather exposes the limits of what can count as 'political' under conditions of coloniality/modernity. As I will show, Aurobindo's turn was still marked by a revolutionary imagination about what was possible for the future of humanity and was therefore resolutely political, but this was a politics which exceeded the narrow confines of what was recognised as politics by the colonial state.

Chapter Seven focusses on the complex international revolutionary networks of the inter-war years through the life of M.P.T. Acharya, a member of the Pondicherry Gang who travelled Europe seeking to overthrow imperialism wherever he found it, and which eventually led him to turn to anarchism as the, to him, ultimate form of emancipatory politics. This chapter exposes some of the spatially extensive networks which were being produced by anticolonialism in the early twentieth century, which not only utilised existing imperial networks but also created new revolutionary spaces by which an alternative world could be imagined. Acharya's biography not only tells us about the spatial extent of anticolonial activism that emerged from the Pondicherry Gang but also con- tinues to challenge the idea that anticolonialism, particularly in the South Asian/ Indian context, was predominantly nationalist in character. This forces us to consider the varied geographies which anticolonial activists were attempting to produce – especially given the tendency today to imagine that independence in the form of a nation-state was the only intended goal of anticolonial movements.

The varied spaces discussed in these chapters, each of which is associated with a particular individual tied in some way to the Pondicherry 'Gang', illustrate how anticolonialism was and is productive of radically alternative ideas about not only the future world after colonialism but also how to get there. One aspect of this which will appear in the book is the debate about whether violent or non-violent strategies were appropriate to defeat colonial rule. For the members of the Pondicherry Gang, they invariably chose the former, at least during the time period covered in this book. A variety of other tensions are also present within any political practice, and anticolonialism is no exception.

Before the next chapter, it is worth discussing my positionality and how this impacts the book. As a white British (Welsh) man, the writing of this book presented a number of challenges. As mentioned in the acknowledgements, I first started research on Pondicherry as a result of a British Association of South Asian Studies/European Consortium for Asian Field Study/British Academy (BASAS/ECAF/BA) fellowship to the Ecole Francais d'Extreme Orient in 2013. However, my contact with Indian Tamil culture had begun when I spent six months between 2003 and 2004 teaching non-formal education in a Tamil- speaking informal settlement in Bangalore (now Bengaluru). However, despite my attempts since then, my Tamil language skills remain very limited – and so I have been reliant in this book upon the work done by a range of Tamil historians of India and, what is today, the state of Tamil Nadu. Foremost amongst these has been Professor A.R. Venkatachalapathy, whose work and also advice throughout this project has proved invaluable. The work done by these academics has proved important in moving sources away from the colonial archive, with all its problems of representation and misunderstanding. As a result, I have always tried to use a variety of archival and other sources, including pamphlets, short stories and car- toons written by the individuals in this book as well as their associates. This, I feel, gives a sense of the rich world which the Pondicherry Gang were creating, and it is due to those Tamil and Tamil-speaking scholars who I am indebted.

Related to this, it should be clear that this is a book primarily written for academic geographers. I hope that it attracts an audience beyond these disci- plinary boundaries. This is important because the story of anticolonialism in South India deserves to be more widely known in general – the paucity of work on South India remains a lack in a number of disciplines, not only geography. However, when I refer to 'we' in this book, I am predominantly referring to geog- raphers. Given the discipline's geographical limits, I have assumed that most people who read this book will be academics based in institutions in EuroAmerica, but also geographers in institutions overseas, not least in India. I sincerely hope that a more international audience reads the book, and think that most of the arguments I make in it would apply equally and make equal sense everywhere – one of the key aims for the idea of anticolonialism I articulate here is a scepticism to disciplinary and subdisciplinary boundaries. However, despite all of this, I recognise that there will be errors and slippages in the text – the responsibility for them, always, lies with me. The cautionary note aside, I now move to discuss what the geographies of anticolonialism could possibly be.

References

Ahmad, A. (1992) *In Theory: Classes, Nations, Literatures*. London: Verso.

Ahmad, A. (1995) 'The politics of literary postcoloniality', *Race & Class*. 36(3), pp. 1–20. doi: 10.1177/030639689503600301.

Ashcroft, B., Griffiths, G. and Tiffin, H. (2000) *Post-Colonial Studies: The Key Concepts*. Oxon: Routledge.

Barker, A. J. (2012) 'Already Occupied: Indigenous Peoples, Settler Colonialism and the Occupy Movements in North America', *Social Movement Studies*. 11(3–4), pp. 327–334. doi: 10.1080/14742837.2012.708922.

Bhambra, G. K. (2014) 'Postcolonial and decolonial dialogues', *Postcolonial Studies*. 17(2), pp. 115–121. doi: 10.1080/13688790.2014.966414.

Bhambra, G. K., Gebrial, D. and Nisanclogu, K. (eds) (2018) *Decolonising the University*. London: Pluto.

Blaut, J. M. (1993) *The colonizer's model of the world : geographical diffusionism and Eurocentric history*. New York: Guilford Press.

Blunt, A. and Rose, G. (eds) (1994) *Writing women and space: Colonial and postcolonial geographies*. New York: Guilford Press.

Blunt, A. and Wills, J. (2000) *Dissident geographies : an introduction to radical ideas and practice*. London: Pearson.

Coulthard, G. S. (2014) *Red Skin, White Masks*. Minneapolis: University of Minnesota Press.

Dabashi, H. (2015) *Can Non-Europeans Think?* London: Zed Books.

Dirlik, A. (1999) 'How the grinch hijacked radicalism: Further thoughts on the postcolonial', *Postcolonial Studies*, 2(2), pp. 149–163. doi: 10.1080/13688799989724.

Esson, J., Noxolo, P., Baxter, R., Daley, P. and Byron, M. (2017) 'The 2017 RGS-IBG chair's theme: decolonising geographical knowledges, or reproducing coloniality?' *Area*. 49(3), pp. 384–388. doi: 10.1111/area.12371.

Fanon, F. (1963) *The Wretched of The Earth*. London: Penguin.

Fanon, F. (1964) *Toward the African Revolution*. New York: Grove Press.

Featherstone, D. (2015) 'Maritime labour and subaltern geographies of internationalism: Black internationalist seafarers' organising in the interwar period', *Political Geography*, 49, pp. 7–16. doi: 10.1016/j.polgeo.2015.08.004

Gahman, L. (2016) 'White Settler Society as Monster: Rural Southeast Kansas, Ancestral Osage (Wah-Zha-Zhi) Territories, and the Violence of Forgetting', *Antipode*. 48(2), pp. 314–335. doi: 10.1111/anti.12177.

Gilmartin, M. (2009) 'Border thinking: Rossport, Shell and the political geographies of a gas pipeline', *Political Geography.* 28(5), pp. 274–282. doi: 10.1016/J.POLGEO.2009.07.006.

Holloway, J. (2002) *Change the World Without Taking Power.* London: Pluto.

Jackson, M. (2014) 'Composing postcolonial geographies: Postconstructivism, ecology and overcoming ontologies of critique', *Singapore Journal of Tropical Geography* 35(1), pp. 72–87. doi: 10.1111/sjtg.12052.

Jazeel, T. and Legg, S. (eds) (2019) *Subaltern Geographies.* Athens, GA: University of Georgia Press.

Kaiwar, V. (2015) *The Postcolonial Orient: The Politics of Difference and the Project of Provincialising Europe.* Chicago: Haymarket.

Kearns, G. (2010) 'Geography, geopolitics and Empire', *Transactions of the Institute of British Geographers.* 35(2), pp. 187–203. Doi: 10.1111/j.1475-5661.2009.00375.x

Legg, S. (2017) 'Decolonialism', *Transactions of the Institute of British Geographers.* 42(3), pp. 345–348. doi: 10.1111/tran.12203.

Loomba, A. (1998) *Colonialism-postcolonialism.* Abingdon: Routledge.

Lugones, M. (2010) 'Toward a Decolonial Feminism', *Hypatia.* 25(4), pp. 742–759.

McEwan, C. and Blunt, A. (eds) (2002) *Postcolonial Geographies.* London: Continuum.

Mignolo, W. D. (2000) *Local Histories/Global Designs: Coloniality, Subaltern Knowledges and Border Thinking.* Oxford: Princeton University Press.

Mignolo, W. D. (2011) *The Darker Side of Western Modernity: Global Futures, Decolonial options.* London: Duke University Press.

Naylor, L., Daigle, M., Zaragocin, S., Ramirez, M.M., and Gilmartin, M. (2018) 'Interventions: Bringing the decolonial to political geography', *Political Geography.* 66, pp. 199–209. doi: 10.1016/J.POLGEO.2017.11.002.

Phillips, R. (1997) *Mapping Men and Empire: a Geography of Adventure.* London: Routledge.

Power, M., Mohan, G. and Mercer, C. (2006) 'Postcolonial geographies of development: Introduction', *Singapore Journal of Tropical Geography.* 27(3), pp. 231–234. doi: 10.1111/j.1467-9493.2006.00259.x.

Quijano, A. (2007) 'Coloniality and Modernity/Rationality', *Cultural Studies.* 21(2–3), pp. 168–178. doi: 10.1080/09502380601164353.

Radcliffe, S. A. (1997) 'Different heroes: genealogies of postcolonial geographies', *Environment and Planning A*, 29(8), pp. 1331–1333. doi: 10.1068/a291331

Radcliffe, S. A. (2005) 'Development and geography: towards a postcolonial development geography?' *Progress in Human Geography.* 29(3), pp. 291–298. doi: 10.1191/0309132505ph548pr.

Rhodes Must Fall Oxford (2018) *Rhodes Must Fall.* London: Zed Books.

Robinson, J. (2003) 'Postcolonialising geography: tactics and pitfalls', *Singapore Journal of Tropical Geography*, 24(3), pp. 273–289. doi: 10.1111/1467-9493.00159

Robinson, J. (2006) *Ordinary cities : between modernity and development.* Abingdon: Routledge.

Robinson, J. (2016) 'Thinking cities through elsewhere', *Progress in Human Geography.* 40(1), pp. 3–29. doi: 10.1177/0309132515598025.

Said, E. W. (2003) *Orientalism.* 5th edn. London: Penguin.

Sharp, J. (2011) 'A subaltern critical geopolitics of the war on terror: Postcolonial security in Tanzania', *Geoforum*, 42(3), pp. 297–305. doi: 10.1016/j.geoforum.2011.04.005.

Sharp, J. P. (2009) *Geographies of Postcolonialism.* London: Sage.

Sharp, J. P. (2013) 'Geopolitics at the margins? Reconsidering genealogies of critical geopolitics', *Political Geography*, 37, pp. 20–29. doi: 10.1016/j.polgeo.2013.04.006.

Sidaway, J. (2002) 'Postcolonial Geographies: Survey-Explore-Review', in Blunt, A. and McEwan, C. (eds) Postcolonial Geographies. London: *Continuum*, pp. 11–28.

Slater, D. (2004) *Geopolitics and the Post-Colonial*. Oxford: Blackwell.

Smith, L. T. (1999) *Decolonizing Methodologies*. London: Zed Books.

Sparke, M. (2005) *In the Space of Theory*. Minneapolis: University of Minnesota Press.

Spivak, G. C. (2003) *Death of a Discipline*. New York: Columbia University Press.

Tuck, E. and Yang, K. W. (2012) 'Decolonization is not a metaphor', *Decolonization: Indigeneity, Education & Society*, 1(1), pp. 1–40.

Wainwright, J. (2012) *Geopiracy: Oaxaca, militant empiricism, and geographical thought*. Basingstoke: Palgrave Macmillan.

Young, R. J. C. (2015) Empire, Colony, Postcolony. Oxford: Wiley-Blackwell.

Chapter Two
Theorising Anticolonial Space

Introduction

As the introduction showed, the emergence of postcolonial thinking in geography from the mid-1980s has opened, not unproblematically, discussions of the intersections between colonialism/imperialism and space. However, as the introduction also showed, these debates are very much alive and contested. A key factor in the discussion of post and decolonial approaches to geography (and other disciplines) is the recognition that imperialistic and colonialist framings of the world inherently create unjust or unequal socio-spatial hierarchies and that these should be resisted (and indeed are) – both postcolonial and decolonial approaches are anticolonial. In this chapter, I wish to think about what shape and form resistance to colonialism takes, and how this 'anticolonialism' has, historically, produced much of the framework for the more contemporary debates in geography. However, I also wish to argue that, more than just being a historical artefact which we can use to construct a genealogy of anticolonial thought/praxis, this approach also offers something different to the postcolonial and decolonial approaches outlined in the introduction. To be sure, the three prefixes 'post', 'de', and 'anti' are all interrelated and share many commonalities, but the anticolonial's emphasis on the political, particularly on both the ideological and material means to resist colonialism, marks it out as distinct.

Throughout the rest of this chapter, I set out exactly what an anticolonial approach to geography could be. This marks something of a turn away from the

Geographies of Anticolonialism: Political Networks Across and Beyond South India, c. 1900–1930, First Edition. Andrew Davies.

more historical material which forms the majority of this book, but is necessary to provide the theoretical framework of the book. The next section discusses how colonialism and anticolonialism have been philosophically proposed as opposi-tional relations of resistance. In particular, I argue that, rather than a 'pure' or reactive form of resistance, anticolonialism should be read as a diverse and pro-ductive form of political activity. This reading relies upon challenging some of the ways in which political concepts do their work. Drawing upon work emerging in minor theory, but also in postcolonial and poststructural readings which have begun to challenge the coloniality of concept-work, I argue that a key point of an anticolonial position is to be sceptical of categorisations and instead to work away at how political boundaries are exceeded and crossed. To begin this process, I examine a 'classic' anticolonial text – Gandhi's *Hind Swaraj*. This section not only helps to ground some of this discussion in a South Asian context but also shows one way in which anticolonialism has been established, and also how the anticolonialisms of the Pondicherry 'Gang' were radically different. To emphasise how this difference worked, I then move on to discuss the ways in which anarchist-inflected geographies and histories have opened up ground to see how solidarities across difference could be built in anticolonial situations. In this section, I draw upon the work of Leela Gandhi and her work on anticolonial forms of solidarity, in particular through her book *Affective Communities*. A concluding section then acts as a transition from this more theoretical reflection on anticolonialism into more specifically South Asian contexts.

Philosophies of Anticolonialism

Anticolonialism is, by its very nature, a mode of political resistance. Howard Caygill (2013) has argued that anticolonialism forms one of a range of political formations which can be classified as 'resistant subjectivities'. To him, these resistant subjectiv-ities 'do not enjoy the freedom of possibility, but only a bare capacity to resist enmity and chance' (p. 98). This, it would seem, positions anticolonialism as a reactive force, that is constituted only because of the colonial dominator's subjugation of the colonised, and means that an anticolonial political subjectivity is only made up of *ressentiment*, or the psychological state driven by deep-seated feelings of hatred or envy. However, drawing from Friedrich Nietzsche and Carl von Clausewitz, Caygill recognises that resistance and *ressentiment* are not simply *reactions to* the political conditions present in any given scenario. In *The Genealogy of Morals*, Nietzsche (1887) outlines the difference between the supposed 'noble' or 'master' form of morality and the 'slave revolt' of morals. The former are forms of morality drawn from 'master' groups in society which creates forms of moral authority through see-ing things that are harmful to themselves as 'bad'. The subjects of these noble forms of morality therefore seek to determine themselves through their difference to these other 'bad' forms of morality. Noble morality/ies produce dominant moral codings and forms which are then imposed on others (the similarities with colonialist forms arguments about civilised behaviours are clear here), and thus 'noble' moral frameworks are active in exceeding their subjects' self-identities and determining what is 'correct' moral behaviour. In contrast, 'slave revolt' moralities are produced from, at the outset, the rejection of what is outside the self, or the attempted imposition of external moral frames, and are predicated upon a desire for revenge against this imposition, which is the core of a position of *ressentiment*.

Given the coloniser/colonised Manichaeism at the heart of colonial rule (Memmi 1974), it is tempting to see this as a suitable way to understand the anticolonial urge to resist. However, the important distinction here is that whilst the desire to resist that which is 'outside' is an important beginning to this process, it is not the sum total of it. As Caygill shows in his discussion of the difference between Marx's and Nietzsche's reading of the Paris Commune of 1871, this lacks a real understanding of how resistance cannot function in a pure morality where one side is productive of a morality whilst another rejects it as 'outside' or 'different'. To Caygill, Nietzsche read the Commune as a Socialist *ressentiment* to the noble morals of the ruling classes, whereas to Marx, both sides were a composite – the reaction of the ruling classes is also a *ressentiment* to the 'new' morality of socialism as it was fought for in the nineteenth century. Seizing the initiative, taking chances and thinking beyond the current 'rules' of the powerful all mean that resistance exceeds being purely reactive, and this is what Marx (1871), in his more affirmative reading of the Paris Commune, termed the 'expansive political form' of new and dynamic modes of political rule. To Caygill then, anticolonial thought must be read in this way, not as a somehow pure form of revenge-desiring *ressentiment* but rather as a potentially expansive and dynamic mode of politics. Caygill draws this out through a reading of Fanon's opening chapter of *Les Damnes de la Terre* (*The Wretched of the Earth*) (Fanon 1963), but here it is useful to think through how this framing of anticolonialism and resistance reads it as a straightforwardly 'Political' activity where resistance is either pre-figured, for example as a necessary action to lead to the future one wants (i.e. a socialist or decolonised future), or is read too simplistically as an almost unthinking reaction of *ressentiment* to a dichotomous world. This creates a quite narrow range of activities or practices that can be thought of as 'anticolonial politics'. The next section moves on to discuss how it is also useful to think through anticolonialism using work that challenges the strictly defined limits of 'the political'.

Minor/Anti/'Small-p' Politics

Debates in the political sciences and political geography have often centred over recent years on the ways in which the political is determined and delimited, and how this in turn creates opportunities for exceptional forms of politics which are then used to govern or discipline society. Foundational to this has been the work of

Giorgio Agamben (1998, 2005), whose work on how 'exceptional' political spaces, such as the concentration camp, create zones of what he terms 'bare life' where human bodies are stripped of their sovereign status and are deemed to be expendable or punishable. This has the political effect of creating a group of subjectivities who are sovereign and whose rights are which have the relational effect of telling those bodies that are deemed sovereign that they are being made secure by this act of separation – at the core of Agamben's work is the political creation and maintenance of the compartmentalisation of life that to him is a modern construct, and the concomitant belief that such practices are from part of the modern narrative of securitisation as a key mode of political control, especially post-9/11. This been used by geographers to explore how certain types of bodies are deemed to be expendable or punishable (Ramadan 2013; Minca 2016; Davies and Isakjee 2018).

However, bare life, securitisation and the disciplinary space of the camp are not simply something that emerged in the twentieth century – many colonial scholars have pointed out the disciplinary and compartmentalising nature of colonisation, most often based around a separation of the coloniser and the colonised (Fanon 1963; Memmi 1974; Pierce and Rao 2006). As Ann Laura Stoler (n.d.) states:

> Colony and camp make up a conjoined conceptual matrix, twin formations and formulations of how imperial, rather than national, logics of security operate. They borrow and blend features of their protective architecture and anticipatory fear. They are in a deadly embrace from the start.

The various examples of colonial rule/discipline which are discussed in relation to the practises of anticolonial resistance in this book show how the anticipatory frameworks of governance and security were written into colonial rule from the outset. However, Stoler's writing is of further use here as it sees the colony as a concept in Deleuzo–Guattarian (Deleuze and Guattari 1987; Deleuze and Guattari 1994) terms – that is, as a constellation or centre of 'vibration', where the various ideas of what a colony is coalesce and become meaningful. For example, it is important to remember that conceptually, a colony is not a single type of socio-spatial arrangement – it could be a farming colony, a slave plantation, or an outpost in deep space – witness the TV series *Battlestar Galactica's* '12 Colonies' of humanity. The question then is what does these various imaginations do? What happens as the particular imaginings of the colony, or any other concept, travel and are deployed ontologically as practices which are not necessarily Political with a capital P? It is clear that all colonies were contingent and provisional constructs – there was never a 'one-size-fits-all' model of constructing and maintaining a colony. Here, it is worth quoting Stoler at length:

> <u>*A colony* does not announce itself as a political concept but exerts its force nonetheless</u>. It organizes visions, imaginaries, and futures. It neither stands alone

nor exists complete, unto itself. It is a potential concept, completed only by what attaches to it, what is excluded from it, and what it attaches to. It is always relational, measured against and distinct from a broader and more stable normative physical, political, and social space: distinct from the normative conventions of "free" settlement, and from a normal population (of those not exiled, ex-communicated, diseased, politically contagious, socially unfit, vagrant or otherwise dis-abled). *A colony* as a common noun is a place where people are moved in and out, a place of livid, hopeful, desperate, and violent – willed and unwilled – circulation. It is marked by unsettledness and regulated, policed migration. *A colony* as a political concept is not a place but a principle of managed mobilities, mobilizing and immobilizing populations, dislocating and relocating peoples according to a set of changing rules and hierarchies that orders social kinds: those eligible for recruitment, for subsidized or forced resettlement, for extreme deprivation or privilege, prioritized residence or confinement. This does not occur in a given, fixed designated space (though claims to rightful sovereignty over a particular place as in the case of colonial conquest over new and otherwise "unproductive" territories and alleged "wasteland" would suggest otherwise). That space is not demarcated once and for all. Its borders shift as do the means to secure them. Some were holding pens and returned to that function in later times; some took as their goal the making of new social kinds sculpted from those cajoled, seduced, chosen or forced to be there. (Emphasis in italics Stoler's, underlined emphasis mine)

As Stoler determines here, the concept of the colony is not formally *P*olitical in that it does not seem to require a formal categorisation as the imaginaries discussed previously show, but it is immanently *p*olitical – the colony in practice shapes a certain range of social and political activities, many of which are geared towards political ends of managing relations of difference. Stoler's work fits into a broader trend of work which has critiqued and challenged many of the assumptions of traditional political philosophy, and which post/anti/decolonial thought contributes to – Gandhi's suspicions of the framings of 'politics' according to Western/Enlightenment values will be discussed shortly, but we could broadly think about such traditional accounts of 'big P' politics as forming part of the 'colonial matrix of power' identified by Walter Mignolo and others (Mignolo 2011). This also aligns with Deleuzo-Guattarian inspired 'rhizomatic' readings of political activity, which exceed, and intermingle across, the boundaries which their opposite, the 'arborescent', seeks to order or control (Deleuze and Guattari 1987). By attempting to define and compartmentalise what is or isn't 'Political', most often taking Carl Schmitt's legally inflected ideas about the political as a realm of intense contestation between 'friend' and 'enemy' (Schmitt 1932), the domain of the political in much political philosophy excludes the 'small p' politics that concepts like 'the colony' shape around them. Anticolonialism, as a political concept, seeks to resist the dominant framings of colonial morals and standards – but crucially it exceeds a Nietzschian state of *ressentiment* as it goes beyond merely being oppositional and instead calls for a radically 'other' imagination of what is possible. Take for example Robbie Shilliam's (2015) *The Black Pacific,* where the counter-narratives of anticolonialism are produced through engagements between black and indigenous Polynesian activists. For Shilliam, European colonial knowledge and science with its belief in the divide between nature and society, fractured indigenous peoples from their lands. However, the domain of the spiritual, whilst affected by colonial knowledge systems, does have areas which were relatively less affected by this – and therefore offers up a space by which indigenous peoples can build what he calls 'deep relation', whereby colonised peoples can relate to each other to find commonalities and use these to 'know themselves better by knowing each other' (p. 14). This is a space of political solidarity, shaped by (post)colonial relationality, but is not about acting or practising certain repertoires or tropes of political activity. Instead, it is the practices that are shaped by the encounter between different groups of colonised peoples and their grounded, lived, responses to both coloniality and their sense of mutuality that is important and crucially productive of different political orders to dominant ones. This also helps to think about how categories like the spiritual can be mobilised to agitate against and decolonise the norms of modernity, which is important for later chapters.

This political position of seeking to work beyond established frameworks is not exclusive to an anticolonial mindset, but it is closely linked to poststructural and postfoundational accounts of politics. Most recently, and from a non-colonial perspective, Emily Apter (2018) has argued that we need to take 'unexceptional' politics seriously. This approach takes seriously the contention that politics is embedded within the everyday, and argues that, rather than a rarified *P*olitical space of true politics, and an exceptional space of political activity drawn from Agamben, there is also a *p*olitical space where traditional limits to political activity are challenged and new concepts, or at least concepts that are newly treated as *p*olitical, are utilised. Apter's book is written for a very contemporary moment, but its premise is useful in thinking through the limits to colonial and anticolonial thought as it is often expressed as will become clear later in this chapter. This also has resonances with work on minor theory in geography, where a series of papers (Gerlach and Jellis 2015; Barry 2017; Jellis and Gerlach 2017; Secor and Linz 2017) has recently reinvigorated Cindi Katz's (1996) call for theories that do not seek to inscribe dominant narratives or framings. There are also resonances with recent work in postcolonial geography, such as Tariq Jazeel's work on translations of geographical knowledge and the importance of taking singularity seriously (Jazeel 2014, 2018). Crucially, we need to understand how the *concept* of anticolonialism mobilised a range of ideas and practises which 'travelled' and were utilised by intellectuals and revolutionaries and the like, but also how they impacted on 'popular' imaginations according to the specific contexts in which they emerged. This is one reason why I have not sought to define exactly what 'colonialism' is or should be, and therefore what anticolonialism is or should be – rather, we should think about anticolonialism as a concept that brings together certain dispositions and practices that were spatially, socially and culturally contingent but can be seen to be 'anticolonial' in orientation. As mentioned in the previous chapter, Stephen Legg has suggested that colonialism as a term is suggestive of 'practices within colonies' (Legg 2017, p. 347). However, this reading of the concept occludes some of the fact that political decisions about colonies and colonisation were always made outside the territorial borders of the colony itself. Likewise, to think about the concept of anticolonialism in this way would determine that it was a series of activities intent on resisting these practises within colonies, which is very much misleading. As this book will go on to show, anticolonialism is as spatially diverse and widespread as colonialism and is always exceeding the category of the colonial. Thus, anticolonial resistance not only draws upon political practises of resistance but also draws in other forms and concepts in order to make its claims. To make sense of this, the next section of this chapter begins to ground some of this discussion in more specifically South Asian contexts through a discussion of one of the foundational texts of Indian anticolonialism, Gandhi's *Hind Swaraj*.

Gandhian Anticolonialism – *Hind Swaraj*

Mohandas Gandhi's volume *Hind Swaraj* (Gandhi 2009) remains recognised as one of his, and one of anticolonial theory's, most important texts. Written whilst travelling from the UK to South Africa aboard the SS *Kildonan Castle*, *Hind Swaraj* was written as a dialogic argument for a form of non-violent anticolonialism. The text itself was only one of a variety of anticolonial texts that were being produced both in South Asian contexts and in anticolonial struggles elsewhere. As will become clear in later chapters, some of these were very closely related to *Hind Swaraj*. Most obviously, Gandhi was writing explicitly against the 'Extremist' wing of the Indian National Congress (INC), which, particularly after the partition of Bengal in 1905, had encouraged violent revolutionary tactics as a way to challenge British imperialism. In particular, Gandhi was writing against the infamous triumvirate of Lala Lajpat Rai, Bal Gangadhar Tilak and Bipin Chandra Pal or 'Lal Bal Pal' who were the figureheads of the extremists, as well as other figures like Aurobindo Ghose. As Gandhi (2009, p. 7) states in the Preface to the English translation of Hind Swaraj which was published in 1909,

> [The extremists] believe that they should adopt modern civilisation and modern methods of violence to drive out the English [from India]. *Hind Swaraj* has been written in order to show that, if they would but revert to their own glorious civilisation, either the English would adopt the latter and become Indianised or find their occupation in India gone.

However, as the aforementioned quote would also indicate, Gandhi was not in line with the other main bloc of the INC, the so-called 'Moderates' who argued for more traditional modes of political contestation against British rule, such as petitioning and gradual reforms leading to a form of self-rule or dominion status for India similar to Australia or Canada. Gandhi's writing in *Hind Swaraj* then marked him out as a maverick thinker of anticolonialism, rejecting orthodox narratives of anticolonialism as nationalist self-realisation.

The key issue, and what made *Hind Swaraj* so influential, was that, to Gandhi, both sides of the INC, and the wider emergent nationalist movement in India, clung too closely to ideas that Western modernity or 'civilisation' as he termed it could provide the answer to India's colonial subjugation, or indeed to human fulfilment more generally. Modernity and industrialisation are the primary target of Gandhi's ire. By the time he was writing *Hind Swaraj*, he had lived in India, the UK, and in South Africa, and in the latter had begun his political activity. In *Hind Swaraj*, Gandhi makes it clear that colonialism is a product of Western modernity and as such can never be wholly removed or defeated by the tools that the ideological systems of the West provide. Gandhi's solution is to blend aspects of 'Western' thought which he found useful, and drawing from the influence the Tolstoy, Ruskin and others had upon him, alongside Indian philosophies drawn from the *Bhagavad Gita* and elsewhere. Gandhi's rejection of modernity then is not a rejection of the West *per se*, but rather a call for a reassessment of the dominant values of the West. As Guha (2013) has noted, Gandhi was, somewhat surprisingly, inspired by G.K. Chesterton, who had written in the *Illustrated London News* a piece based upon his readings of and conversations with the various Indian nationalists based in London which was sceptical of the originality of Indian nationalism. As a result, Gandhi saw the need to create a less derivative proposal for what an 'independent' India could be. This reassessment involved challenging some of the fundamental classifications and categorisations of the West, including the division between the personal and the political, but also using Indian philosophy as the foundation of this new politics, and it is worth spending some time to go through exactly how Gandhi did this.

The very terminology of the book's title represents a challenge to these orthodoxies. Using *Hind*, an alternative and vernacular term for the geographic territory of India, represents a (relatively standard) rebuttal of colonialist arguments that India was not a contiguous territory before the onset of colonial rule. However, *Swaraj*, literally meaning 'self-rule' (swa or sva = the self, raj = rule/authority), as deployed by Gandhi explicitly blends both the political and the personal. To Gandhi, the idea that politics could somehow be seen as a separate realm from the moral self was anathema. Swaraj was carefully developed to include a double meaning in Gandhian terms. Whilst it could mean political independence for India, or 'home rule' as the INC termed it at the time, it also crucially meant the governance of the self. Indeed, in the original Gujarati, the term Swaraj was used interchangeably for both of these terms, and it was only in the English translation published two months later that Gandhi alternated between Swaraj and home rule in describing the two different aspects of personal/

political (Parel 2009). This conceptual challenge to a distinct realm of 'the political' is crucial to Gandhian thought, both in this relatively early stage and after. It also leads Gandhi towards some of the more controversial passages of the book – where, most notably in Chapter Five: 'What is Swaraj' he argues that simply removing the 'English' from India rather than creating Hindustan would rather create an 'Englistan', where the rules and norms of Western civilisation are maintained with no foreseeable shift in the attitudes or practices for the vast majority of the Indian population. Swaraj, to Gandhi, is what happens 'when we learn to rule ourselves' (Gandhi 2009, p. 71), and this rule is as much personal and moral as it is formally 'political'. The foundation for this comes from Indian philosophy, whereby the 'aims of life' or *purusharthas* of ethical integrity (*dharma*), wealth and the political (*artha*), pleasure (*karma*), and spiritual excellence (*moksha*) are all relatively autonomous parts of the whole system. These aims of life, according to Gandhi in Hind Swaraj, should not only be the aims of the individual but also the community, the nation and indeed, all civilisations. As Parel (2009) notes, this was strikingly innovative, as Indian philosophy had either focussed on, or was perceived to focus on, *moksha* at the expense of *artha*. To Gandhi, the bringing of these four aspects into balance was also not a purely Indian question but was rather a question for the world – India could teach the world about these issues, and this is about more than simply pasting the need for spirituality, *moksha*, transcendence, or any other stereotype as the total solution and endgame of the imagined future after colonial rule ends, it is instead about fundamentally reassessing the values of Western modernity and how they had been unevenly spread around the world through colonial rule, and the toolkit that Indian philosophy could provide to address that.

Swaraj was a term in use before Gandhi's intervention, but it forms a key aspect of the wider Gandhian philosophy of *satyagraha*. *Satya* (Truth) and *graha* (firmness/insistence/to hold fast) formed the core message of specifically Gandhian modalities of non-violent resistance. Meaning more than simply 'passive resistance', *satyagraha* is a commitment to thinking, speaking and acting in ways which are 'truthful', which to Gandhi meant a commitment to acting in ways which were non-violent. Swaraj is central to this doctrine, as the commitment to the truth firstly comes from the self. Thus, key to Gandhi and his wider political philosophy is the role of self-management or self-governance, starting with the individual and then expanding outwards, most famously to the village level, but also to the wider nation and the world beyond India. This aspect of moral self-rule is also central to many anticolonial ethical positions but also extends beyond them – as Leela Gandhi (2007) has argued, the moral framework of *satyagraha* as self-rule was both a rebuttal to colonial rule, but also provided an outlet for utopian socialisms, anarchisms, and more recently environmental political movements.

This leads on to the second aspect of Mohandas Gandhi's thought in *Hind Swaraj* which is worth detailing here – the importance of non-violence. Obviously, non-violence is now a central pillar of how we read and make sense of Gandhi in the twenty-first century the world over, but *Hind Swaraj* is where he articulated his ideas at some length formally for the first notable time. Gandhi felt driven to this by the events he had witnessed in London before he boarded the *Kildonan Castle*. The most infamous spur was the assassination of the India Office official Sir William Curzon Wyllie by Madan Lal Dhingra at a meeting at the Imperial Institute in London on 1 July 1909. Dhingra was a member of the extremist faction of India revolutionaries based at India House in Highgate and his act was loudly supported by leading members of the group, such as V.D. Savarkar (Chaturvedi 2013). At the same time, extremists in India had been writing about the different varieties of resistance – Aurobindo Ghose published a series of articles under the title 'The Doctrine of Passive Resistance' in *Bande Mataram* in April 1907. To Aurobindo, passive resistance was one tool to move towards independence for India, but violence was another which was acceptable in certain circumstances (Ghose 2002; Heehs 2008; Mahajan 2013). The open violence being espoused by members of the India House group, and the rhetoric of extremists in India, meant that the case for non-violence needed to be more clearly made. As Gandhi stated, in a passage seemingly speaking directly to the extremists:

> Whom do you suppose to free by assassination? The millions of India do not desire it. Those who are intoxicated by the wretched modern civilization think these things. Those who will rise to power by murder will certainly not make the nation happy. Those who believe that India has gained by Dhingra's act and other such acts in India make a serious mistake. Dhingra was a patriot, but his love was blind. He gave his body in a wrong way; its ultimate result can only be mischievous. (Gandhi 1909 [2009], pp. 75–76)

Finally, it is also worth reflecting upon the fact that *Hind Swaraj* is also important as a geographical text of anticolonialism. To start with, it is a profoundly inter/trans-national text. By the time of its writing, Gandhi's thought had been shaped by life not only in India but also his political campaigning in South Africa and his life in London. The insistence of Gandhi on challenging the idea that nationalism, independence or any other form of 'home-rule' that relied purely upon Western notions of statehood or government is challenged throughout the text by examples Gandhi draws from both his travels but also his knowledge of the world. Thus, when, during the book's dialogue, 'The Reader' challenges the Gandhi-proxy 'Editor' about following the Italian path towards nationhood, Gandhi argues back that whilst figures like Mazzini and Garibaldi are 'adorable' (p. 73), the emergence of the Italian nation-state after the *risorgimiento* has had very little material benefit to the working classes. This line of argument is a clear rebuttal to many of the extremists' lines of argument, where individuals like Mazzini could be held up as a figurehead of successful national reintegration. However, strategically, in *Hind Swaraj*, Gandhi downplayed this internationalist element and almost entirely ignores his agitations in South Africa, in order to emphasise the 'Indian' nature of Swaraj as a distinct future civilisation (Hyslop 2011).

It is no surprise that both *Hind Swaraj* and Gandhi's later writing have been taken up by a number of writers since. *Hind Swaraj* in particular received a degree of criticism from both the colonial authorities – where unsurprisingly the book was banned by the Government of India (GoI) in March 1910 – but also amongst a variety of Indian readers – it was castigated by Shyamji Krishna Varma, the founder of India House, in *The Indian Sociologist* in 1913 (see Parel, p. lxxi). The banning of the book in India meant that it did not achieve widespread traction or discussion until after 1919, where it was seen as the manifesto for Gandhianism. Extremist opinions could be broadly read as similar to Krishna Varma's reaction in 1913, although much of the revolutionary zeal of the 'extremist' INC in 1909 had been diluted and replaced by different tactics in the intervening years. Unsurprisingly, Indian Marxists and Communists like M.N. Roy (1922) and S. Dange (1921) criticised Gandhi for not recognising the class elements that were holding Indians back from prosperity – for them he overemphasised the humanitarian urge. From the more moderate perspective, Sir Sankaran Nair in his *Gandhi and Anarchy* (Nair 1922) argued that Swaraj as Gandhi outlined it was practically unworkable and would lead to complete chaos. Of continued importance to India lies the fundamental disagreement that Gandhi had with B.R. Ambedkar. Ambedkar's status as a Dalit leader and his opposition to the caste system meant he fundamentally disagreed with many of Gandhi's proclamations about the suitability of Indian religion and civilization to promote moral goodness within the wider world. Ambedkar wrote little about *Hind Swaraj* directly, but his position can be judged accurately from their other disagreements, and the blind spot that the Gandhi had for Dalit political consciousness at the expense of orthodox Hindu teachings continues to form a core space of antagonism against the internal colonialisms of Indian society in contemporary Dalit-Bahujan and decolonial politics today (Kambon 2018).

However, the political method and approach that Gandhi began to enunciate clearly in *Hind Swaraj* continues to be a crucial aspect of both his political philosophy and how he is read today. Often, these writers have focussed on how Gandhi's specific strategy or mode of resistance *satyagraha* (truth-firmness, or a commitment to the truth), a technique he began developing in South Africa and then went on to deploy in India – was the result of his thinking about anticolonialism. This is unsurprising given the role that Gandhi plays in subjects like peace studies and related areas as well as the impact that Gandhian thought has had on movements in the present (Williams and McConnell 2011; McConnell 2014). *Satyagraha* also provides us with an understanding of how Gandhi sought to combine and work through the challenge of becoming a resistant subject discussed earlier. *Satyagraha*, often mistranslated as passive resistance, was instead a refusal to engage with the colonial oppressor on their terms. Instead of violence, non-cooperation, and the firm moral commitment to these, even up to the point of death, was key. As Caygill (2013) points out, to Gandhi, the key moment by which one became a *Satyagrahin* was the taking of an oath, and this moment transforms the subjectivity of the anticolonialist. It is this transformation between a dominated subject of a colonial power and instead into an active and non-conforming resistant subjectivity that exists outside the framing of colonialism that is the key element of Gandhian thought here.

Gandhi's importance to anticolonial thought, especially in the South Asian context is undoubted. However, just as important here is the sense by which this discussion of Gandhian politics through the prism of *Hind Swaraj* indicates the varied and contested world of anticolonialism which was present in the early twentieth century. The target of *Hind Swaraj* was undoubtedly the kind of people who formed the Pondicherry Gang – extremist 'nationalists' who sought to desta-bilise the ruling colonial order through whatever means necessary. To Gandhi, a key aim was to challenge the supposed ideas of modern development tied to Western ideas of civilisation which many of the extremists supposedly had. Whilst this is undoubtedly true of some of these men, the later chapters of this book will also show how Gandhi's position represented only one part of the range of extremist and revolutionary positions which were at work within the emergent Indian independence movement. Gandhi's *Hind Swaraj* then must be read as being produced at a particular moment – for Gandhi, one that is resolutely placed in the South African- and London-based imperial spaces which he was, at that time, most embedded in.

The later chapters of this book make it clear that South India around the time of Gandhi's writing offered a very different set of possibilities about how to ima-gine and contest colonial rule. Some of these would have been relatively familiar to Gandhi, for example, the international revolutionary networks which various anticolonialists passed through – such as M.P.T. Acharya's various moves through European-based radicalism which forms the focus of Chapter Seven. However, others would not have, and indeed the South Indian context of anticolonial organising has hardly been examined in any real depth – something which is explored throughout this book. However, this is much more than simply recog-nising that there was a regional spatiality to anticolonialism – a return towards an old fashioned 'regional geography' of political activity. Instead, it is to recognise that anticolonialism is much more varied than outright displays of resistance (whether violent or non-violent) but is also productive of new sociopolitical forms. In this regard, we must therefore see that Gandhi and *Hind Swaraj* were, for many anticolonialists, neither foundational nor essential, nor did it adequately represent the world that they either inhabited or wished to create. The following chapters of this book suggest how the Pondicherry Gang was intimately connected with a range of different spaces of anticolonialism – from the creation of shipping lines, the establishment of urban practices/repertoires of protest (McAdam,

Tarrow and Tilly 2001; Tilly and Wood 2012), the shaping of emergent Indian Communism, as well as the spaces of the spiritual 'reawakening' of India to name only a few.

This examination of Gandhian thought through the lens of *Hind Swaraj* has highlighted some of the limits of anticolonialism as only ever produced as a form of *ressentiment,* and he remains profoundly important reference point for anticolonial thought. We could also include other thinkers like Frantz Fanon in this way, recognising that Fanon's anticolonialism was routed through the particular experiences and structures of French Algeria and the French Caribbean, but also through the international context of decolonization which marked the 1950s and 1960s (Fanon 1963; Macey 2012). Yet, whilst Gandhi (and Fanon) both provide important conceptual frameworks for thinking anticolonially, it is clear that there are a variety of ways of envisioning what shape and form anticolonialisms may take, and it is important to move beyond these two authors or other 'key' figures to avoid reifying them as the only thinkers of anticolonialism. If we are following Stoler's advice to think about colonialism as a constellation or alignment of principles, then we need to examine some of the other ways in which this arrangement has been challenged, and in the final substantive section of this chapter, I introduce some of the ways in which geographers and others have begun to interrogate anticolonialism.

A Further Critique of Anticolonialism as Nationalism or What Is Anticolonialism?

As the previous discussions have shown, much anticolonial thought bordered a number of related areas and was often not only 'nationalist' in character. Instead, the circuits of anticolonialism, especially in the first half of the twentieth century were occupied by a variety of intellectual positions. Whilst the focus on Gandhi has allowed a sense of some of the dominant trajectories of anticolonialist thought, thinking about other anticolonialisms is important in showing the very real diversity of anticolonialism which is often elided. This foregrounds the different imaginaries of anticolonialism which were at work and which have proved a fruitful space for geographical scholarship in recent years. Whilst, as Kearns (2014) has noted, not all anticolonial thought is creative and imaginary, I want to stress here and throughout the rest of this book is exactly the very resourceful and creative responses to colonialism which were produced by individuals, often in situations of hardship and duress. Thus, whilst nationalism is undoubtedly important, it was not the be all and end all of anticolonial intellectual and political endeavour. Thinking beyond nationalism is particularly important in some of the earlier (i.e. pre-World War One) anticolonialisms, which, I would argue, must be read intersectionally.

Over the past decade or more, an emerging trend in studying anarchist histories and geographies has usefully worked at the interstices of anticolonial thought (Springer et al. 2012; Maxwell and Craib 2015). More specifically to the intellectual history of geography, Federico Ferretti (Ferretti 2013, 2016, 2017a, 2017b) in particular has shown over a number of studies of anarchist intellectuals like Elise Reclus and others that internationalist networks created dynamic counter-currents to imperialist and capitalist thought. Similarly, Benedict Anderson's *Under Three Flags* (2007) took the writings of the Philippine writers Jose Rizal and Isobelo de los Reyes and inserts them into a wider trajectory of dissident organising that stretches across the globe. Anderson shows how, even in the colonial space of the Phillipines, the emerging European nation-state world-empire emerging after Bismarck's unification of Germany, the politics and dreams of the international left in the wake of the Paris Commune of 1871, and the slowly declining world of the Spanish Empire all showed different 'worlds' to people like Rizal and Reyes. This moves supposedly 'marginal' or 'peripheral' colonial spaces like the Philippines to become more central in the long struggle against imperialism and the emergence of distinct anticolonialisms. This is an important geographical work in continuing to challenge and presumed pre-eminence of certain areas of the world. It is also important in the context of this book, which seeks to push back against the established position that groups like the Pondicherry 'Gang' were only ever marginal sideshows in the grand sweep of anticolonial activity in the twentieth century.

This internationalist and anarchist trend within anticolonialism is most clearly visible in Maia Ramnath's (2011a) *Decolonising Anarchism*, which excavates the often forgotten intersections between anarchism and anticolonial movements in the most sustained study of this type so far. Challenging ideas that see anticolonialism as only rooted in nationalism or nation-state centric thought, Ramnath shows how reading South Asian anticolonialism with an anarchist lens uncovers different aspects of the nationalist struggle and reveals that anarchism was always present with 'mainstream' anticolonialism's trends. This is also important as the anarchist refusal of the state can also de-couple anticolonial struggle from the temporal moment of decolonisation – the granting of independence. As Ramnath states,

> [S]tandard nationalist history tells one story of decolonization. There are others, and they are still unfolding. In these stories, the achievement of a national state was not the endpoint of liberation, and its inherited institutions not the proper vehicle. The elimination of the British Government left incomplete the task of ending injustice and iniquity. The postcolonial state, insufficient at best, at its worst actually perpetuated the same kinds of oppression and exploitation carried out by colonial rule, but now in the name of the nation. (Ramnath 2011a, pp. 4–5)

The various discussions about the nature of the freedom struggle and after in South Asia will be discussed in the next chapter, but important here is the sense in which the anticolonial is a broader and more diverse set of practices than it is often assumed to be. As the discussion of Gandhi has previously shown, many were sceptical of the ability of the nation-state to provide ultimate freedom from colonial rule. For instance, Ramnath has used this approach to explore other specific moments in India's freedom struggle, such as the Ghadar movement's attempts to foment an internationalist rebellion (Ramnath 2011b) or Dalit-Bahujan struggles against the colonialist state in post-independence India (Ramnath 2015).

However, this anarchist tendency is also something that can be seen as deeply rooted in the ethical imperative underlying anticolonial thought. In a series of interventions into postcolonial theory, Leela Gandhi has broadened the boundaries of anticolonial thought and challenged various critiques of the political nature of anticolonialism. Firstly, in a response to critiques of postcolonial thought from both the orthodox Marxist critiques outlined in the introduction and from more recent poststructuralist (neo)Marxist arguments about the (seemingly lacking) political nature of postcolonialism, Gandhi (2011) links the aspects of the minor theory outlined previously, anarchist reactions against formal politics and the anticolonial in productive ways. Anticolonial contexts vary, but at least in many of the resistances to European colonialisms, and the narrative of 'the civilising mission' that many deployed, it is relatively clear to see that a scepticism towards organised and institutionalised political forms, most of which could be allied to the colonial/imperial state, would become part of many anticolonial movements. This leads to many of the claims about the political naivety of much anti-, de-, and postcolonial thought – it fails to deal with the truly *Political* at the expense of thinking about the mundane, the moral, and the everyday. This is certainly a reason why the Nehruvian postcolonial government of the Republic of India abandoned some aspects of Gandhian *satyagraha* as impractical and idealistic rather than practical and properly political (see also Gandhi 2007, 2014 here). Anticolonial ethics then, following Leela Gandhi, are inherently unruly, and whilst they did not always adopt the anarcho-libertarian/communist principles of mutual aid and solidarity (see for example people like V.D. Savarkar or the various other anticolonial movements and intellectuals that adopted quasi-Fascist or other less than progressive political frameworks), the anarchic impulse can be seen as an originary part of many of these movements.

Gandhi has most fully developed her position further in *Affective Communities* (2006). Rooted in Derridean post-structuralism, alongside postfoundational conceptions of the political, Gandhi's argument is diverse and wide ranging and is worth spending some time detailing. To her, anticolonialism in the years around the turn of the nineteenth and twentieth centuries was heterodox and diverse, drawing in a number of different social reformers and activists, some of whom could or would not have been considered 'political' at the time. However, it is the intersectional and affective nature of connections between anticolonialists, both 'Western' and 'colonised', that forms a crucial space in which to challenge imperialism.

Gandhi's point in *Affective Communities* is to stress this intermediate zone as it comes to exist between anticolonialists who live in the spaces of the coloniser and the colonised – not only rupturing the already porous boundary between the two but also recognising that claims to hybridity and fluidity are problematic. To her, the hybrid subject of postmodernity is too driven by approaching the world as a source of self-enjoyment – whereas the Kantian or Marxist agent:

> arrives into self-regard through ascetic self-enclosure, the hybrid subject reaches a similar destination through an insatiable demand for self-fulfilment, consuming the very world in/for/from which it must fashion its ethical capacity... Polymorphous and perverse, the hybrid subject is cloned, we might say, from the genetic substance of corporate capital and the world market. (p. 22)

So, to Gandhi, we require a different set of theoretical tools to understand how subjectivity can be built. That moves beyond the binaries of modernity and postmodernity, which helps to understand the complex relationship that anticolonialists like Mohandas Gandhi had with modernity. To achieve this, Gandhi utilises Derrida's politics of friendship – where friendship is a more-than-political act which challenges the boundaries between individuals and groups that colonialism hoped to establish – Gandhi sets out to uncover the 'friendships' between western anticolonialists who were both sympathetic to, and often formed crucial parts of anticolonial activities. This approach, given Gandhi's suspicion of the hybrid subject, moves beyond the broadly postcolonial perspectives of the likes of Said (2003, 1993) and Bhabha (1994), who, in their emphasis on the contrapuntal/hybrid/interstitial provide much of postcolonial thought with its suspicion of the Manichean binarisms (such as core/periphery, coloniser/colonised, metropolitan/colonial) through which imperialism sought, and failed, to create its own version of order. However, unlike Said and Bhabha who look back towards the failures of imperialism to create and sustain binary differences amongst the various different subjects in their critique, Gandhi points out how the boundaries were challenged both by anticolonial nationalists, but also by 'metropolitan' elites. As Gandhi avows, one of the key ways in which friendships between these binaries could be created and shaped was through things like the vegetarianism (in the case of Mohandas Gandhi and Henry Salt) or, as we shall see later in this book, the spiritual actions of Mirra Alfassa and Sri Aurobindo. So, if these friendships cannot be understood through either a modernist Kantian or Marxist perspective, or a postmodern hybrid subject, what is the alternative pathway?

Gandhi's answer to this binarism comes through an engagement with Jean-Luc Nancy's work on community and the idea of being in common (Nancy et al. 1991; Nancy 1992). Here, relationality between two individuals is built through sharing and 'compearance' between two individuals. Compearance is the process by which two individual agents can come together and discover mutually common ground, without giving up their sense of self. This approach means that rather than by building political identities through subsuming oneself into a community that demand aspects of uniformity as the 'price' of joining (e.g. one's race, class, gender) and inevitably failing in seeking some form of 'perfect' community, instead one can find and build solidarities that instead think of a more utopian, always incomplete, and perpetually deferred futures. This approach does not magically dissolve some of the underlying ethical conundrums of the individual's obligations (or lack of them) to a wider group, and it is here that Gandhi argues that we instead of individuality, we should think in terms of singularity, what Olivier Marchart (2007) has termed the 'being-in-common of singularity' (see also Jazeel 2018). Here it is necessary to turn briefly to the idea of how this approach challenges conventional notions of the political.

This postfoundational approach is rooted in the very obviousness of much of the 'political' – or what, in Heideggerian terms could be thought of as the 'retreat of the political' – not in that the political ceases to be, but rather that it is so obvious that *everything is political*, that we become oblivious to its taking place around us at all times. As a result, for Nancy, the political needs to be retraced and reworked in new ways that make the mundane nature of much political activity visible. This version of the political is not an autonomous domain that can be rendered distinct where only certain actions are deemed to be political. Thus, to Gandhi, one of the key components of this 'co-belonging of nonidentical singularities' (Gandhi 2006, p. 26) is that it challenges and resists the state's attempt to clarify what a properly political community is. Given the imperial and colonial state's often rampant desire to classify and categorise its subjects, alongside the current state's governmental and developmentalist impulses, the postfoundational approach advocated by Gandhi provides useful tools by which to think about anticolonialism as a realm of possibility that challenges conventional orthodoxies. Thus, whilst colonial subjects were confronted by the very stark realities of foreign rule on a regular basis, there were numerous ways in which the overall system of imperial/colonial politics functioned that were taken for granted, and others in which they were resisted that were so commonplace, or were so different to the colonial typology of what counted as 'political', that they were largely ignored or excluded from the political record.

This postfoundational politics does useful work in beginning to destabilise what counts as the properly political, but also helps to move past some of the binary tendencies between elite and subaltern practices of resistance which have dominated South Asian historiography since Ranajit Guha's (1982) opening treatise on the necessity for subaltern knowledge to be discussed in the archive and which will be discussed in the next chapter. Gandhi's focus on the negotiation of difference between anticolonial 'Western' and 'non-Western' subjects overlaps somewhat with the deep relation between indigenous communities of Robbie Shilliam, but crucially, neither authors' ideas seek to overcode or universalise the discrete anticolonialisms which these encounters produce. This approach lies at the heart of this book's aims, and I utilise these ideas to worry away at the actual practices and spaces of anticolonialism as they existed in the past, and, which can be used equally importantly, as they continue today. This, similar to Ramnath's arguments about anarchism which were discussed earlier, does important work in stretching both the boundaries of imperialism/colonialism and resistance to it – what I term here the geographies of anticolonialism.

Conclusions

The aforementioned discussions of anticolonial thought are not intended as providing an idealised or correct version of anticolonial thought/praxis. Instead, I utilise Leela Gandhi's arguments about the politics of friendship to emphasise how anticolonialism involves a commitment to openness and solidarity to ones fellow colonised subjects. However, thinking geographically about this position importantly spatialises these somewhat lofty ideals and extends them beyond simply a vague tolerance for difference to see how they actually operated in practice. The ways in which the different spatial forms that later chapters address (maritime and more-than-landed, urban but culturally distinctive, networked and international, amongst others) will illustrate what an anticolonial approach offers to the understanding of politics and geography. It is worth stressing again here that this is not intended to claim that the anticolonial is somehow 'better' than decolonial or postcolonial approaches. Rather, thinking anticolonially brings into focus a different set of frames and repertoires that operate alongside and with these other theoretical approaches. Many of the authors cited previously would probably identify as postcolonial or decolonial scholars first, rather than 'anticolonialists' *per se*. However, by focussing on how these authors have thought through the politics of anticolonialism, this usefully shows how resistance to colonialism produces a politics that exceeds its supposed limits. Anticolonial politics, as Mohandas Gandhi and others like Fanon recognised, was always more than simply a *ressentiment*, but thinking anticolonially exposes how this political activity necessarily forces us to reconceptualise 'the political'.

For the rest of this book, I use this anticolonial 'concept' as outlined previously to think through the varied anticolonial geographies that were produced through a specific set of relations that converged around Southern India in the early twentieth century. In the next chapter, I go through some of the historiographical and geographical debates about anticolonialism in South Asia more generally, as well as introducing some context about South India in preparation for the historical chapters that follow. These chapters each focus on specific instances, moments, places or individuals and are all quite closely linked, but each produced radically different outcomes. Some are inherently nationalist, whilst others end with a wholesale rejection of the *P*olitical. This is also important because it shows the range, scope and vitality of anticolonial thought.

References

Agamben, G. (1998). *Homo Sacer: Sovereign Power and Bare Life*. Stanford: Stanford University Press.

Agamben, G. (2005). *State of Exception*. Chicago: University of Chicago Press.

Anderson, B. (2007). *Under Three Flags: Anarchism and the Anti-Colonial Imagination*, 1e. London: Verso.

Apter, E. (2018). *Unexceptional Politics: Obstruction, Impasse and the Impolitic*. London: Verso.

Barry, A. (2017). Minor political geographies. *Environment and Planning D: Society and Space* 35 (4): 589–592. https://doi.org/10.1177/0263775817710089.

Bhabha, H.K. (1994). *The Location of Culture*. Oxford: Routledge.

Caygill, H. (2013). *On Resistance: A Philosophy of Defiance*. London: Bloomsbury.

Chaturvedi, V. (2013). A revolutionary's biography: the case of V D Savarkar. *Postcolonial Studies* 16 (2): 124–139. https://doi.org/10.1080/13688790.2013.823257.

Dange, S.A. (1921). *Gandhi vs. Lenin*. Bombay: Liberty.

Davies, T. and Isakjee, A. (2018). Ruins of empire: refugees, race and the postcolonial geographies of European migrant camps. *Geoforum* https://doi.org/10.1016/J. GEOFORUM.2018.09.031.

Deleuze, G. and Guattari, F. (1987). *A Thousand Plateaus*. London: Continuum.

Deleuze, G. and Guattari, F. (1994). *What Is Philosophy?* London: Verso.

Fanon, F. (1963). *The Wretched of The Earth*. London: Penguin.

Ferretti, F. (2013). "They have the right to throw us out": Élisée Reclus' New Universal Geography. *Antipode* 45 (5): 1337–1355. https://doi.org/10.1111/anti.12006.

Ferretti, F. (2016). Geographies of peace and the teaching of internationalism: Marie-Thérèse Maurette and Paul Dupuy in the Geneva International School (1924–1948). *Transactions of the Institute of British Geographers* 41 (4): 570–584. https://doi.org/10.1111/tran.12143.

Ferretti, F. (2017a). Publishing anarchism: Pyotr Kropotkin and British print cultures, 1876–1917. *Journal of Historical Geography* 57: 17–27. https://doi.org/10.1016/J.JHG.2017.04.006.

Ferretti, F. (2017b). Tropicality, the unruly Atlantic and social utopias: the French explorer Henri Coudreau (1859–1899). *Singapore Journal of Tropical Geography* 38 (3): 332–349. https://doi.org/10.1111/sjtg.12209.

Gandhi, L. (2006). *Affective Communities: Anticolonial thought, Fin-de-Siecle Radicalism, and the Politics of Friendship*. London: Duke University Press.

Gandhi, L. (2007). Postcolonial theory and the crisis of European man. *Postcolonial Studies* 10 (1): 93–110. https://doi.org/10.1080/13688790601153180.

Gandhi, L. (2011). The Pauper's gift: postcolonial theory and the newdemocratic dispensation. *Public Culture* 23 (1): 27–38. https://doi.org/10.1215/08992363-2010-013.

Gandhi, L. (2014). *The Common Cause: Postcolonial Ethics and the Practice of Democracy, 1900–1955*. London: University of Chicago Press.

Gandhi, M.K. (1909 [2009]). *Hind Swaraj and Other Writings* (ed. A.J. Parel). Cambridge: Cambridge University Press.

Gerlach, J. and Jellis, T. (2015). Guattari: impractical philosophy. *Dialogues in Human Geography* 5 (2): 131–148. https://doi.org/10.1177/2043820615587787.

Ghose, A. (2002). *Bande Mataram: Political Writings and Speeches 1890–1908. Complete Works of Sri Aurobindo, Vols 6 and 7.* Pondicherry: Sri Aurobindo Ashram.

Guha, R. (1982). On some aspects of the historiography of colonial India. In: *Subaltern Studies (Volume 1) Writings on South Asian History and Society.*

Guha, R. (2013). *Gandhi Before India.* London: Penguin.

Heehs, P. (2008). *The Lives of Sri Aurobindo.* New York: Columbia University Press.

Hyslop, J. (2011). An "eventful" history of Hind Swaraj: Gandhi between the Battle of Tsushima and the Union of South Africa. *Public Culture* 23 (2): 299–319. https://doi.org/10.1215/08992363-1162048.

Jazeel, T. (2014). Subaltern geographies: geographical knowledge and postcolonial strategy. *Singapore Journal of Tropical Geography* 35 (1): 88–102.

Jazeel, T. (2018). Singularity. A manifesto for incomparable geographies. *Singapore Journal of Tropical Geography* https://doi.org/10.1111/sjtg.12265.

Jellis, T. and Gerlach, J. (2017). Micropolitics and the minor. *Environment and Planning D: Society and Space* 35 (4): 563–567. https://doi.org/10.1177/0263775817718013.

Kambon, O. (2018). The Pro-Indo-Aryan Anti-Black M.K. Gandhi and Ghana's #GandhimMustFall Movement. In: *Rhodes Must Fall: The Struggle to Decolonise the Racist Heart of Empire* (eds. R. Chantiluke, B. Kwoba and A. Nkopo), 186–206. London: Zed Books.

Katz, C. (1996). Towards minor theory. *Environment and Planning D: Society and Space* 14 (4): 487–499. https://doi.org/10.1068/d140487.

Kearns, G. (2014). "Up to the sun and down to the centre": the utopian moment in anticolonial nationalism. *Historical Geography* 42: 130–151.

Legg, S. (2017). Decolonialism. *Transactions of the Institute of British Geographers* 42 (3): 345–348. https://doi.org/10.1111/tran.12203.

Macey, D. (2012). *Frantz Fanon: A Biography,* 2e. London: Verso.

Mahajan, G. (2013). *India: Political Ideas and the Making of a Democratic Discourse.* London: Zed Books.

Marchart, O. (2007). *Post-foundational Political Thought: Political Difference in Nancy, Lefort, Badiou and Laclau.* Edinburgh: Edinburgh University Press.

Marx, K. (1871). *The Civil War in France* (eds. B. Baggings and M. Carmody). Marxists Internet Archive.

Maxwell, B. and Craib, R. (eds.) (2015). *No Gods, No Masters, No Peripheries: Global Anarchisms.* Oakland: PM Press.

McAdam, D., Tarrow, S., and Tilly, C. (2001). *Dynamics of Contention.* Cambridge: Cambridge University Press.

McConnell, F. (2014). Contextualizing and politicizing peace: geographies of Tibetan satyagraha. In: *Geographies of Peace* (eds. N. Megoran, F. McConnell and P. Williams), 131–150. London: I.B. Taurus.

Memmi, A. (1974). *The Colonizer and the Colonized.* London: Earthscan.

Mignolo, W.D. (2011). *The Darker Side of Western Modernity: Global Futures, Decolonial Options.* London: Duke University Press.

Minca, C. (2016). *Agamben and Geography.* London: I.B. Taurus.

Nair, C.S. (1922). *Gandhi and Anarchy.* Madras: Tagore.

Nancy, J.-L. (1992). La comparution/the compearance: from the existence of "Communism" to the community of "Existence". *Political Theory* 20 (3): 371–398. https://doi.org/10.1177/0090591792020003001.

Nancy, J.-L. et al. (1991). *The Inoperative Community*. Minneapolis: University of Minnesota Press.

Nietzsche, F. (1887). *On the Genealogy of Morals*, 1998e. Cambridge: Hackett.

Parel, A.J. (2009). *Editor's Introduction to the Centenary Edition'*, in Gandhi, M.K. (1909 [2009]) 'Hind Swaraj and other writings, Edited by Parel. A.J., Cambridge: Cambridge University Press.

Pierce, S. and Rao, A. (eds.) (2006). *Discipline and the Other Body*. Durham, NC: Duke University Press.

Ramadan, A. (2013). Spatialising the refugee camp. *Transactions of the Institute of British Geographers* 38 (1): 65–77. https://doi.org/10.1111/j.1475-5661.2012.00509.x.

Ramnath, M. (2011a). *Decolonizing Anarchism: An Antiauthoritarian History of India's Liberation Struggle, Anarchist Interventions*. Edinburgh: AK Press.

Ramnath, M. (2011b). *Haj to Utopia: How the Ghadar Movement Charted Global Radicalism and Attempted to Overthrow the British Empire*. Berkeley: University of California Press.

Ramnath, M. (2015). No gods, no masters, no brahmins: an anarchist inquiry on caste, race and indigeneity in India. In: *No Gods, No Masters, No Peripheries: Global Anarchisms* (eds. B. Maxwell and R. Craib), 44–79. Oakland: PM Press.

Roy, M.N. (1922). *India in Transition*. Geneva: J.B. Target.

Said, E.W. (2003). *Orientalism*, 5e. London: Penguin.

Said, E.W. (1993). *Culture and Imperialism*. London: Chatto and Windus.

Schmitt, C. (1932). *The Concept of the Political*. 1996 trans. (ed. G.D. Schwab). Chicago: University of Chicago Press.

Secor, A. and Linz, J. (2017). Becoming minor. *Environment and Planning D: Society and Space* 35 (4): 568–573. https://doi.org/10.1177/0263775817710075.

Shilliam, R. (2015). *The Black Pacific: Anti-colonial Struggles and Oceanic Connections*. London: Bloomsbury.

Springer, S. et al. (2012). Reanimating anarchist geographies: a new burst of colour. *Antipode* 44 (5): 1591–1604. https://doi.org/10.1111/j.1467-8330.2012.01038.x.

Stoler, A.L. (n.d.). Colony. *Political Concepts* (1) http://www.politicalconcepts.org/issue1/colony/.

Tilly, C. and Wood, L.J. (2012). *Social Movements: 1768–2012*, 3e. London: Routledge.

Williams, P. and McConnell, F. (2011). Critical geographies of peace. *Antipode* 43 (4): 927–931. https://doi.org/10.1111/j.1467-8330.2011.00913.x.

Chapter Three
South India and Anticolonialism: The Minor Politics of Anticolonialism in a Historiographical 'Backwater'

Introduction

The history of India's movement for independence from British rule is hardly a subject that is under-examined, with thousands of studies of the various aspects of the emergence and development of what eventually became the Republic of India post-1947. This chapter provides an overview of these debates, but focusses most on those which are of most relevance to the content of this book. The majority of this book's later historical sections deal with events that occurred in the so-called pre-Gandhian phase of the freedom struggle, prior to Gandhi's return to India from South Africa in 1915. As Chandra (2012) has argued, this pre-Gandhian phase was vital in producing an ideology of anticolonial struggle, even if it did not acquire the mass-movement characteristics that defined the Gandhian phase.

Chandra's typology of this pre-Gandhian phase provides a useful framework to build the later discussions of this chapter around, as he argues that there were three distinct aspects of pre-Gandhian anticolonial nationalism. Firstly, the development of the idea of India as a coherent 'nation', or at least nation in becoming, contra to colonialist claims that it was not, and never could be. Secondly, the increasing prominence of the 'drain theory', whereby India was seen to be being drained of its wealth by British imperialism. Developed by moderate and liberal thinkers like Dadabhai Naoroji, and increasingly used by more extremist thinkers like Lala Lajpat Rai, drain theory provided a powerful

Geographies of Anticolonialism: Political Networks Across and Beyond South India, c. 1900–1930, First Edition. Andrew Davies.
© 2020 Royal Geographical Society (with the Institute of British Geographers). Published 2020 by John Wiley & Sons Ltd.

counterpoint to the colonial authorities' proclamations of benevolence, but was also a strong bonding element as it exposed both colonial exploitation and was so self-evident that it did not have to be explained through complicated arguments (Chandra 1965). Finally, the nationalist agitations at this point deployed the language of democracy and civil liberties, utilising these concepts, often drawn from 'Western' enlightenment thought, in order to challenge the colonial authorities. Particularly evident here was the notion of freedom of the press which the emergent nationalist and often vernacular (i.e. written in the local language, whether Bengali, Punjabi, Tamil or any other South Asian language) press was able to take advantage of.

Whilst Chandra's typology is broadly accurate, it also relies upon a conception of the freedom struggle being motivated purely along nationalist lines, and as if the nation-state of the Republic of India was the only possible, and indeed, the only desired, outcome for the majority of those involved in anticolonial agitation. Given that, as Goswami (2012) has argued, revolutionary movements like anticolonialisms are primarily motivated by a desire to (re)shape the potential future, it becomes clearer to see how utopianisms, cosmopolitanisms and other ideas circulated alongside nationalisms within the anticolonial milieu, as was discussed in the previous chapter. As a result, this chapter sets out a number of trends in the established literature which have emerged in scholarship on anticolonial politics in India more generally. These include discussions of violence and non-violence within the freedom struggle, the increasing importance of international studies to understanding imperial (and anticolonial) geographies, and the continuing discussion of the subaltern in relation to these histories and geographies. In addition to this, the chapter discusses and introduces the particular context of South India during the first few decades of the twentieth century. This latter section will both provide some context and background to those who may be unfamiliar with South India but will also begin to trace why exactly the anticolonial geographies which came into being in this region are worthy of greater academic attention.

Violence and Non-violence

The centrality of Gandhian *Satyagraha* (as discussed in the last chapter) to popular narratives, especially in the West, of the struggle for independence in South Asia has often occluded the many ways in which violence was both central to imperial rule, but also to resisting it. This has a number of consequences which are relevant to today, especially when considered in terms of the debates about the benevolence or rapacity of British rule in India, which has become an increasingly polarised debate in recent years, especially in the public sphere (see for example the debates around the publication of Shashi Tharoor's *Inglorious Empire* in Tharoor 2017; Allen 2018; Roy 2018). However, this rendering has remained powerful, partially due to the effects of colonial nostalgia promoting the idea of the UK willingly giving up the 'jewel' in its Empire peaceably, but also thanks to the continued desire for non-violent solutions to seemingly intractable political situations. Gandhi's charismatic appeal as the *Mahatma*[1] and his voluminous amount of writing on non-violence is a key aspect of this. However, it is increasingly recognised that the Gandhian and Indian National Congress (INC)-led mass mobilisations that were non-violent in nature were only one aspect of the struggle for freedom. As Maclean (2015) has explained in her study of the Hindustan Socialist Republican Army (HSRA) in the interwar period, the boundaries between the nominally non-violent Congress and the revolutionary aspects of the freedom struggle were much more blurred than is usually credited, and they often operated in what Maclean calls 'collegial interaction' (p. 7). Maclean's work shows how revolutionary violence remained an important concern for both the Government of India (GoI) and the Congress – and arguably led to a number of distinct policy shifts on both sides of imperial/nationalist struggle – from Gandhi's return to political activity in the Calcutta Congress of 1928 following an outbreak of political violence (Brown 1977) through to changes in GoI policy following the armed revolt of the Royal Indian Navy after World War Two (Davies 2013).

This is not to fetishise violence but rather to recognise that violent activity remained an important aspect of the struggle against imperialism throughout the independence movement. It is important to note at the outset the importance of the violence of the state/empire in attempting to contain violence in both everyday and seemingly mundane ways but also in extreme moments such as the rebellion of 1857 or the Amritsar Massacre of 1919 (Wilson 2016; Wagner 2017). However, outbreaks of anticolonial and resistant violence often occurred even during notionally 'Gandhian' movements/protests, most infamously in the Chauri Chaura incident during the Non-Cooperation Movement in 1922, where a number of police were burned to death in their *chowki* (station) following a clash between protesters and police. Gandhi withdrew from INC political activity as he claimed that the masses were not prepared for independence and in utilising non-violent methods, and the INC halted Non-Cooperation as a direct result. Whilst Chauri Chaura has attracted notable academic discussion, particularly from a subaltern perspective (Amin 1995), outbreaks of mass violence were common throughout the struggle for independence, culminating in the widespread atrocities that occurred in the run up to and during the partition of British India into independent India and Pakistan (Chandra et al. 1989).

More specific were acts of revolutionary violence which were deliberately intended as challenges to colonial rule. The release of the records of the Indian Political Intelligence (IPI) records in the British Library in the 1990s provided a stimulus to much of this work. The IPI was formed in Britain in 1909 as a means

---

[1] 'Mahatma' meaning 'Great Soul' was an honorific applied to Gandhi after 1914 following his struggle against racial injustice in South Africa.

to surveille Indian revolutionaries in London and abroad and was active until India's independence (O'Malley 2008). The release of the archives has allowed an insight into the range and scope of imperial intelligence agencies, and as Brückenhaus (2017) has argued, this often prefigures more recent forms of counter-terrorism as the tactics and strategies of the authorities evolved. This work has, importantly, stretched the temporal framings of revolutionary violence, showing that debates about violence and non-violence continued throughout the entire freedom struggle, whereas it was previously often seen as predominantly occurring in the pre-Gandhian era as part of an early upsurge in activity before nationalist sentiments were fully formed.

However, recent work such as Maclean's discussed previously has increasingly challenged this framing and has importantly re-emphasised how limited colonial archives are in providing details of the undercover and illicit activities which were essential to anticolonial, revolutionary activity, instead turning towards other sources such as newspapers, oral histories, and images (Maclean and Elam 2013; Elam and Moffat 2016; Maclean 2016). There has also been an increased interest in the intellectual work which was done during these times operated at the intersections between violence and non-violence (Bose and Manjapra 2010). For instance, Sri Aurobindo, the subject of Chapter Six, who was one of the key figures of the 'extremist' wing of the INC and was implicated in bombing conspiracies, also wrote extensively in the publications he edited (such as *Bande Mataram,* *Karmayogin,* and *Dharma*) on issues such as what 'freedom' meant in an Indian context (Mahajan 2013; Wolfers 2016). It is also clear that the *Swadeshi* movement (c. 1905–1908), often characterised as the first mass nationalist political movement after the 1857 rebellion, was diverse in its intellectual and political imagination. As Bate (2012, p. 42) has pointed out about the diverse nature of the *Swadeshi* movement, it

> articulated virtually every major idiom that would define the freedom struggle, Indian nationalism, and even postcolonial democratic politics in the 20th century – boycott and the promotion of *swadeshi* commerce, especially in textiles; the appeal to labour; the use of folk motifs in song and story; new literature, poetry and drama in political protest; the production of nationalist space and time; and, of course, the systematic interpellation of the "people" as a new political agency through swadeshi languages.

Whilst the *swadeshi* movement formed only one aspect of this pre-Gandhian 'phase', it is often seen as the hallmark of the developing anticolonial movement at this time, with a diverse but intersecting range of political strategies that were being developed and mobilised during the time. As will become clear in Chapter Four on the activities of V.O.C. Pillai and the *Swadeshi* Steam Navigation Company in the far south of Madras Presidency, *swadeshi* activism often involved violence, from both the state and 'the people', but was about much more than that in its attempts to experiment with and imagine a future free from colonialism.

Whilst the tactics of *swadeshism* remained a part of the freedom struggle up to independence and after, the specific *Swadeshi* Movement of c. 1905–1908 failed as the colonial state reacted, often with violence, to shut down the emergent movement. In some ways, this has been used as an example of how naïve the more revolutionary and extremist elements were – by acting violently with only a small number of people involved these early revolutionaries seemingly highlighted their inability to act *P*olitically correctly. The wider reaction to this failure is also important as it further pushed at the fractures between extremists and moderates which were at play within the INC – Gandhi's *Hind Swaraj* was written after the peak of the swadeshi movement, for example. In addition, Shruti Kapila (2010) has shown how, whilst the 'extremists' were at odds with Gandhian thought, even as it emerged in the 1900s and 1910s, they were formulating intellectual responses to it, and this had fundamental implications for what the 'political' (in a South Asian context) was. To Kapila, it is a mistake to see Gandhi's return to India from South Africa in 1915 as a Hegelian moment of synthesis between the thesis/ antithesis of the moderate/extremist views at work within the INC (for a related discussion of the intersections between Gandhi and the 'extremist' position, see Ghosh 2016). Instead, taking the 'extremist' nationalist Bal Gangadhar Tilak's translation of the *Bhagavad Gita* (which Tilak did whilst imprisoned in Rangoon at around the same time Gandhi was writing *Hind Swaraj*), Kapila argues that Tilak's reading of the political and political violence did not use the 'traditional' (i.e. Western or, more specifically, Schmittian) or even Fanonian understanding of anticolonial conflict occurring between a binary of Friend & Foe.

Instead, the fraternal was the key binding of political solidarity, for both Gandhi and Tilak (and many others in the freedom struggle). This is important as it framed Hindu's and Muslim's, in Tilak's view, as brothers within the same 'household' of India. The core narrative of the *Bhagavad Gita* is written in the form of a discussion between the god Krishna and Prince Arjuna of the Pandava dynasty about Arjuna's dilemma about launching into a war against his cousins in the Kaurava dynasty. To Tilak, this exposed the distinct nature of the political and political subjectivity in India as founded upon forms of internecine and fraternal violence – whilst Krishna and Arjuna's debate in the *Gita* cover many ethical and philosophical topics, the foremost for Tilak is Krishna's attempts to push Arjuna towards declaring war and fulfilling his spiritual duty. For extremists like Tilak, the recognition that violence had seemingly always been a mechanism of the political in South Asia was a realistic and pragmatic one. Gandhi's arrival in India and the widespread utilisation of the concept of *satyagraha* did not replace or efface the possibility of this fraternal violence, as has been proven by the all too common outbreaks of communal violence which have occurred regularly during and since independence/partition.

The failure of the *swadeshi* movement meant that intellectuals like Tilak returned to writing about what an independent, non-colonised, political Indian subject would look like. Given the violence with which the state reacted to swadeshi mobilisations, reading classical Indian texts like the *Gita*, which deal with spiritual and political desires for freedom, provided space for Tilak to imagine a distinctly Indian and modern political subject that brought together aspects of tradition but placed them alongside resolutely modern concerns about nation- and selfhood. Thus, to Kapila, Tilak's (and other extremists') writings at this time expose how the political can be mobilised and shaped through violence in an Indian context. The writings of the extremists at this time need to be recovered and re-examined because, as Kapila puts it of Tilak, '[they] did make violence possible, plausible and conceivable' (p. 440), and this violence has been all too clear in India's polity in both pre- and postcolonial eras.

This turn towards a renewed engagement with aspects of revolutionary violence in the history of Indian independence is important in challenging public stereotypes of the non-violence of India's decolonisation, but also has important consequences for understanding the diverse and contested nature of the 'internal' politics taking place throughout the freedom struggle, but also has important consequences post-independence. Kachwala (2018) has, for example, recently explored the ways in which anticolonial violence has been represented as a masculine act/performance, but also one which was fundamentally necessary to resist the supposedly greater violence of British colonialism, in Indian popular cinema.

Violence is thus an important concern when attempting to understand the movement for Indian independence. As we will see in the later empirical chapters, it was relatively easy for anticolonialists to resort to violence, or for violence to break out during anticolonial movements. This was more than just the violence of 'the mob' or 'rioting' and attempting to understand how and why violence occurred is important to understand the contested geographies of anticolonialism in India.

The National and Inter/Transnationalism

Following from the discussions of violence, it is somewhat unsurprising, given the eventual establishment of the territorial nation-states known as India and Pakistan (and eventually Bangladesh), that the history of the freedom struggle is often conflated with nationalism. Nationalism was, of course, of huge significance in providing political impetus for India's eventual freedom, as well as in carving out the future social and cultural imaginary of 'India' as a distinct polity (Kaviraj 1992). As Chandra (2012) has argued, the nation of India was created through the process of anticolonial struggle. Whilst the geographical and cultural unity of India was somewhat coherent prior to colonisation, the struggle against colonialism meant that a distinct political and economic 'India' had to be created which had never existed in such a way before (Goswami 2004). This process did not occur quickly or seamlessly, with a variety of regional cultural and political processes acting alongside broader 'All India' struggles to shape the nation. Thus, the contested and 'fragmented' way in which this evolved has been the subject of much study (Chatterjee 1993; Thapar, Noorani and Menon 2016), and ongoing debates about the nature of India's unity or otherwise continue to be important.

However, the nation-centric account of India only tells one version of the history of the freedom struggle. The turn towards global and international histories and historical geographies has been important here (see, for example Bayly 2004; Osterhammel 2014; Burton and Ballantyne 2016), as well as studies which interrogated simplistic renderings of the imperial world as spatially divided between core and periphery (Ghosh and Kennedy 2006). For example, the networked accounts of history that emerged after the 2000s chimed strongly with debates about networks and globalisation which were taking place in geography (Lester 2001, 2005). The spatiality of empire then remains a core concern, and geographers have been crucial in exploring the diverse and interconnected geographies which were created and maintained by imperial rule (Legg 2007, 2009, 2014). David Featherstone has also written extensively about the various anticolonial solidarities which were important in shaping the inter- and transnational geographies of resistance (Featherstone 2008, 2012, 2015). My own work elsewhere has utilised a variety of spatial vocabularies, from assemblages to the transnational to understand some of these anticolonial political processes (Davies 2012, 2013, 2019).

This turn towards networked or other extra-territorial history struggles for a vocabulary at times. Is it possible to speak of an 'international' or 'transnational' form of connection between two or more areas which were colonised and therefore were not strictly 'nation-states', and were often made up of multiple 'nations'? Likewise, what does using 'transcolonial' (Ghosh and Kennedy 2006) obscure as much as it reveals in altering the spatial emphasis of these terms? Here, Itty Abraham's discussion of what constitutes international space is particularly useful:

> international space is best understood as a product of struggle: over things, boundaries, membership, movement and passage, rules, and ideas. International space emerged from struggles taking place at multiple locations among and between globally dispersed empires, territorial states, anti-systemic political movements, campaigns for national liberation, transnational corporations seeking new avenues for investment and trade, socially committed civil society, and religious organizations working in different parts of the globe, as well as from the tensions attendant on movements of people travelling long distances for reasons ranging from pilgrimage to indentured labour. (Abraham 2015, p. 5)

This sense of the international emphasises the dynamic interconnections which worked through (and importantly against) the European imperial world order of the early twentieth century. Thus, various international spaces were vital to the functioning of the broader imperial system. However, far from an ordered and martialled system of control, the increasing mobility of individuals across imperial spaces opened up and reworked relationships in multiple ways, from creating new rhythms of life according to the regularity of steamship travel (Anim-Addo

2014) to challenging nation-centric politics through black internationalist movements like Pan-Africanism (Iton 2008; Hodder 2016). Despite the challenges of using these terms, the way in which they allow a detailed exploration of the various connections that occurred across imperial boundaries and between imperial/colonial spaces means that these intermediate 'international' spaces of encounter, dialogue and solidarity are an increasing focus.

Much of the discussion of the 'international' in relation to India focusses on the interwar period. This makes sense given the widespread turn towards politically 'internationalist' organisations after the devastation of World War One, but also because the increasing visibility of anticolonial movements globally during this time allowed greater connections between them to occur. For example, the meeting of the International Congress against Imperialism and Colonialism in Brussels in 1927 (which formed the League Against Imperialism (LAI)) is often argued to prefigure the internationalist and anticolonialist ethos which marked the Bandung Conference of the Non-Aligned Movement in 1954 (Prashad 2008; Petersson 2014).

There is also an important set of discussions taking place about how Indian anticolonialisms intersected with different political movements. Partially emergent from the research on global histories/geographies, this has produced crucial work which has destabilised established nationalist and other dominant framings. For example, Kate O'Malley's work on Irish and Indian anticolonial solidarities (O'Malley 2008) was important in exposing the long standing connections which were built between anticolonialists in the two countries (and which were being observed by British colonial and imperial authorities) which cemented the post-independence politics of the two nations.

The importance of seeing the metropolitan 'centre' as a space of anticolonial resistance and agitation is also important. As Ahmed and Mukherjee's (2012) collection shows there were various examples of diverse political activities which took place in the UK, and in reality, the metropole offered a range of important advantages to the colonised, as the dehumanising laws of the colonies could not be brought into effect in cities like London. The role of gender is also of importance here, especially as inter- and transnational spaces could, potentially, offer ways for dominant gender (and other) hierarchies to be challenged and disrupted. So, for example, Mukherjee (2018) shows how Indian women were significant presences in the organisation of international struggles for suffrage. Indian women were politicised in these struggles but were also subject to the colonial hierarchies of the day – solidarity did not remove or equalise existing differences across social class or race. As a result, whilst the politicisation of Indian women can be read as an example of Gandhi's affective politics of friendship (Gandhi 2006), the spaces of suffrage, similarly to the spaces of anticolonialism, did not automatically create spaces of understanding or solidarity. Instead, and developing on the arguments in the last chapter, political friendship (as in any other form of long-term relation) must be worked at in order to transcend difference and build the being in common of singularity.

As Mukherjee (2018) shows, this process varies across space – with nationalist spaces creating different political arrangements to feminist spaces, for example. This again is important to understand how the anticolonial spaces of South India intersected with other political struggles, and whilst the later examples I draw upon are overwhelmingly very masculine spaces of contestation, gender played an important role, even if it was deployed by the men in this book in often patriarchal ways. One way in which this occurred was through the deployment and increasing development by Indian nationalists of the idea of *Bharat Mata*, or the nation of India as a divine mother goddess, was important here (Ramaswamy 2010). Here, masculinist nationalists often called on other men to defend the honour of India – the mother – from her desecration by the coloniser.

However, the role of gender relations needs to be read in the context of the diverse currents of international cosmopolitanism. Again, Gandhi's work on drawing together friendship across conditions of difference is important here, and especially in the fin-de-siècle of the nineteenth and twentieth centuries, as this was a particularly intense period of utopian cosmopolitan interaction. In the South Indian context, the most common example of this international cosmopolitanism was the Theosophical Society. Founded in 1875 in New York City as a spiritual society which believed in an ancient mystical tradition which had emerged from Tibet and was aimed towards human unity, by 1880, the leaders of the Society, Helena Blavatsky, Colonel Henry Olcott and William Judge, had established the international headquarters in Madras, on the banks of the Adyar river in the south of the city. After Blavatsky's death, the group fragmented, with Judge returning to the United States. In India, Olcott remained, and by 1907, Annie Besant had emerged to lead the Society in Madras. Besant was born in London in 1847 but was a supporter of Irish and Indian independence, as well as campaigning on women's rights. Besant then, to some extent, represents a clear example of the intersections of various radical and utopian traditions. Under Besant's leadership, the Theosophical Society became connected to the INC, and Besant was instrumental in forming the Home Rule League on the outbreak of World War One. Like the *Swadeshi* movement, the Home Rule League was another marker in the growing demand for Indian independence, especially at a time where the British Empire was calling upon its possessions to fight in the war.

The Theosophical Society's engagement with the politics of Indian nationalism largely falls outside the remit of the later discussions of this book. The members of the Pondicherry 'Gang' had either become marginalised or had become affiliated to other political movements by the time of Besant's most radical activities. However, the development of cosmopolitan forms of identity across difference is important to understand how anticolonialism was variously influenced by a variety of spiritual, radical and utopian impulses in the early twentieth century (Mohanty 2015). This continues to extend and challenge the territorially bounded and nation-centric accounts of the emergence of anticolonial thought. Groups like the Theosophical Society were concerned with combatting imperialism and colonialism not to create a new system of nation states but instead for the benefit of all of humanity.

A further important consequence of this variety is, as Raza, Roy, and Zacharia (2015) have argued, understanding how interwar Indian internationalism was an important mechanism in challenging the often prefigured assumptions and classifications of political ideologies which have become cemented with the benefit of hindsight. Thus, it is easy now to categorise certain individuals and their ideas as 'Communist', 'Socialist', 'Fascist', or similar, but, at the time, these terms often did not have the well-defined characteristics, nor carry much of the historical 'baggage' that they do today. Thus, the intersections between and across ideologies, especially ones that were not diametrically opposed, were far less clear. The most common example here is of the international revolutionary V.D. 'Veer' Savarkar. Today, Savarkar is most often recognised as the author of 'Hindutva', or the political position of 'Hindu-ness', which has become synonymous with the chauvinist Hindu-supremacist politics of postcolonial India. However, Savarkar was (and is) a more complex figure than this and was involved in revolutionary networks in London and Paris, where he wrote the anticolonial account of the 1857 Rebellion, *The History of the War of Indian Independence* (Savarkar 1909), which inspired many Indian anticolonialists, including the men in Pondicherry, and was swiftly banned as seditious and was the most notable early attempt to reconfigure what the imperial authorities termed the 'Indian Mutiny' (Chaturvedi 2013). Thus, although talking about specifically interwar instances of internationalism, Raza et al.'s work shows how these later mutable political classifications were based upon previous activities, to which the men of the Pondicherry Gang were closely connected. However, in the same volume as Raza et al., Zachariah (2015) also argues for caution in emphasising the international at the expense of other spatial scales or framings, and it is worth restating that, similar to any spatial framework, the international does not provide a 'perfect' understanding of the imperial/colonial world of the first half of the twentieth century.

These internationalist histories and geographies are important in that they expose various spaces of imperialism and anticolonialism. In the fin-de-siècle era and after, networks of revolutionary anticolonialists and revolutionaries from a number of colonial contexts emerged, which quickly became the subject of surveillance and policing as documented by Brückenhaus (2017) and O'Malley (2008). Harald Fischer-Tine's (2007) survey of the range of internationalist connections between revolutionary Indians in London, New York, and Tokyo is instructive here, but also represents only a sample of the urban centres beyond India which hosted anticolonial activists for short periods of time. Many of these networks were explicitly active in metropolitan rather than colonial areas – deliberately taking advantage of the inability of colonial states to impose laws that limited civil rights and liberties on their 'home' subjects that they could enforce upon their subjects in the colonies (Tickell 2011) – whilst strategies of 'containing' revolutionaries in exile often proved counter-productive when exiles from a variety of colonial contexts were brought together and shared proximate space to discuss and share ideas (Kothari 2011). More diverse again in its coverage of the range of activities conducted by anticolonial movements internationally are Maia Ramnath's detailed studies (Ramnath 2011a, 2011b) of the various transnational movements which were active in this period, and which were noted in the previous chapter. The importance of these international spaces to the establishment of anticolonial political formations is central to understanding how anticolonial narratives and ideologies circulated, but also how they forged solidarities. Thus, thinking back to the idea of anticolonialism as a 'concept', it is vital to think of it as something that extended beyond the boundaries of any one colony or emergent nation-state. We have already seen how *Hind Swaraj* was a transnational text written between London, South Africa and India, but it is increasingly clear, as Maclean and Elam (2013, p. 115) state, is that 'revolutionary thought in late colonial South Asia was the product of and a participant in a global network of Indians', and the diverse, and often contradictory, ideologies and debates that were involved amongst these diverse networks produce an alternative image of the modernising and globalising world that was emerging.

In addition to the spatial extent of these revolutionary networks, it is also worth noting here the methodological challenges necessary to this approach to inter/transnationalist history. Subaltern Studies' impact upon how colonial archives were read (see the next section) and written about meant that 'official' sources of knowledge like the archives of the state were recognised as only one, often limited, source of information, and one which should be treated with a degree of scepticism (see also Stoler 2008). The shortcomings of these archives to understand international Indian revolutionary activity, drawn as they largely are from the partial (and often paranoid) official secret service reports collected by the IPI in London or the Criminal Investigation Department (CID) of the GoI means that they are necessarily limited sources and should be treated as such. One way to address this shortcoming has been to broaden the idea of what is an acceptable historical source is, with studies increasingly incorporating the visual and aesthetic into studies of the emergence of Indian nationalism (Ramaswamy 2010), or including the visual, oral and textual as well alongside official and non-official archival sources (Maclean 2016). It is also increasingly commonplace to read multiple archives together – exploring how events were recorded in multiple colonial locations to piece together fragments of these inter/transcolonial histories (Anderson 2012).

These discussions of internationalism will be visible throughout many of the later chapters. One of the key aspects of the various anticolonial spaces which the 'Pondicherry Gang' and its associates worked through was that it always connected to wider intellectual discussions and networks the extended across the globe. Thus, in later discussions of Subramania Bharati in, we may find the importance of Mazzini's thought, whilst the discussion of M.P.T. Acharya in Chapter Seven will place him as a key member of some of the international revolutionary networks mentioned previously.

The Elite Versus the Subaltern

The last of the major areas of literature related to Indian historiography which I wish to discuss here is the 'subaltern'. The term subaltern is now relatively commonplace within academia, indicating 'persons and groups hierarchically positioned as subordinate or inferiors within nation-states, capitalist production relations, or relations of patriarchy, race, caste, and so forth' (Gidwani 2009, p. 66). However, it is worth going over the history of the term, emerging as it does from a distinctly Indian context to history.

The Subaltern Studies Collective (SSC) was formed by a number of historians of South Asia in the late 1970s. The initial manifesto for the group was set out by Ranajit Guha in 'On Some Aspects of the Historiography of Colonial India' published in the first *Subaltern Studies* volume in 1982. In this short piece, Guha set out to critique the vast majority of existing historical work on India on the grounds that it was elitist – either colonial elitist (i.e. written from the perspective of the coloniser) or bourgeois-nationalist elitist (i.e. written from the perspective of the elites who came to rule and govern India after independence in 1947). This approach adopted a version of Marxism that was heavily indebted to Gramscian Marxism (the usage of the term 'subaltern' in these terms comes from Gramsci's prison notebooks) and to the 'History from Below' approach which was pioneered by the likes of E.P. Thompson in the 1960s. Broadly put, to Guha and others in the SSC, India's independence had simply transferred ruling power to a new ruling bourgeois-nationalist class and had failed to truly provide any meaningful change to the vast majority of India's citizens. Writing, as the likes of Guha were, in the aftermath of Indira Gandhi's 'Emergency' (1975–1977) which curtailed many democratic rights in India, it seemed that the venality and corruption of elite bourgeois-nationalism was clear.

A second important trend in much SSC writing was the publication of Said's *Orientalism*, with its resultant scepticism towards knowledge of the colonial 'other' produced in imperial centres. Thus the other target of the SSC in its early years was the Cambridge School of Indian historiography, which at the time was the most important academic group involved in producing histories of India. The Cambridge School histories of colonial rule were drawn largely from British or imperial sources and often downplayed social and cultural histories at the expense of formal political mechanisms. To the SSC, they also emphasised the benevolent and modernising tendencies of Empire and saw much Indian political manoeuvring as self-serving game-playing rather than as true political behaviour. This manoeuver reified only a certain number of administrative functions as 'political'

activity (Guha 1998), which removed self-representation from the actual partici-
pants of history (O'Hanlon 1985) and acted as a fundamental limit to how the
past could be understood – as Pandian (1995, p. 387) put it: 'there can be no
more histories other than what the Cambridge school designates as history'.

Taking on these established figures meant that the SSC was soon critiqued
from a number of perspectives, and the ideas of the group evolved over the next
decade to become more influenced by Foucaultian and other post-foundational
thought in response. Chaturvedi (2000), in his introduction to a collection of
texts on Subaltern Studies, argues that this shift occurred during 1986 as a result
of critiques of the SSC project from Marxist Indian scholars of peasant rebellion,
a tradition which the collective had not engaged with as a result of the British
Marxist tradition of history from below which they emerged from. Alongside this
was the recognition within the collective that the search for a 'true' or 'essential'
peasant or subaltern consciousness was fruitless, and an increasing awareness of
the various aspects of the collective's foundational thought which could be considered Eurocentric. These together meant a shift away from the traditions of
historical materialism towards more 'poststructural' responses. This approach
reached something of an apogee with Gayatri Chakravorty Spivak's foundational
critique of the idea of the subaltern, voice, and the role of European philosophers
speaking on behalf of others in *Can the Subaltern Speak?* (1988).

The range and scope of publishing by the SSC's members was (and remains)
vast, including: surveys of policing and carcerality by David Arnold (Arnold
1986, 1994); discussions of how political violence was integrated or rejected from
mainstream discourses of the freedom struggle at different times (Amin 1995);
critiques of Eurocentric knowledge (Chakrabarty 2007); studies of nationalism
from non-elite perspectives (Chatterjee 1993); and theorisations of how British
rule was maintained in conditions where it never achieved a formal hegemony
(Guha 1998), to name only a few. The SSC's project remains controversial,
despite the last volume of work, *Subaltern Studies XII*, being published in 2005.
Critiques of many of the intellectuals within the movement have often verged on
the personal – based largely on the fact that many of the members of the collective
have gained employment in US institutions as a result of the project's infamy. Its
authors are still the subject of critique today, such as in Vivek Chibber's (2013)
polemic where he criticises certain key texts written by the group, but groups
them somewhat misleadingly under the title of 'Postcolonial Theory' rather than
Subaltern Studies, and which led to a particularly bitter public debate between
Chibber and Partha Chatterjee which was widely circulated on youtube (and is
available at https://www.youtube.com/watch?v=xbM8HJrxSJ4).

A more balanced critique of the postmodern/poststructuralist nature of some
of the SSC's work, in particular Dipesh Chakrabarty's *Provincialising Europe* and
Ranajit Guha's *Dominance without Hegemony* comes from Kaiwar (2015). To
Kaiwar, particularly in its later incarnations, Subaltern Studies' Heidegger-
inflected poststructuralism means that it ends up missing the core of its original

aims – that of understanding and challenging inequality. In Marxist terms, late-era Subaltern Studies, at least to Kaiwar, has become obsessed with the super-structure, and forgotten about the base – works like Chakrabarty's, whilst arguing for a recognition and tolerance of difference, cannot provide more than that, as he puts it '[r]especting the beggar's right to sleep under a bridge or on the sidewalk, and his/her invocation of the gods and spirits in a cruel world, while avoiding such historicist issues as income redistribution is in fact to give the game away' (p. 165). This Marxist critique chimes with the critiques of Dirlik and Ahmad in the introduction, but I will return to this critique in the later chapter on Aurobindo Ghose. Equally importantly for this book, Kaiwar also recognises the slippage in many Subaltern Studies texts written by Bengalis which position Bengali cultural and political practises as synonymous with Indian practises. One of the crucial points of this text as a whole is to provide another geographical lens to understand the emergence of anticolonial politics in regions other than Bengal which tend to be under-recognised.

Another critique of Subaltern Studies links more closely to the ideas of violence discussed previously. The SSC's emergence around the time of both 'The Emergency' but also the Maoist inspired Naxalite rebellions in West Bengal (which continue in India's Eastern and Southern states today) led to various members of the group, most notably Ranajit Guha, to relatively uncritically endorse violent rebellion as a method. As Hardiman (2013) has shown, this was problematic in at least two ways. Firstly, the role of violence in the later *Hindutva* mobilisations of the 1980s and 1990s exposed the limits of this approach, and an increased recognition that some forms of subaltern militancy were retrograde and non-progressive. Secondly, this also exposes the tension within the early SSC work which drew upon Marx's critique of 'passive resistance' as a method by which the bourgeois could counter revolutionary militancy. As a result, Guha (1992) could claim that Gandhi, and Gandhian non-violence, was against the masses, something that became clear when the façade slipped and Gandhi referred to the people as a violent 'mob'. As Hardiman argues, this does a disservice to the ideas of Gandhi, but also means that we should examine the intersections of anticolonial violence with gender and other socio-political categories. As Hardiman (2013, p. 44) states:

> Violence as a method is most suited to able-bodied males, with women, the elderly and the very young unable, as a rule, to play much part. The need for arms and training similarly excludes many. Violence is either the method preferred by small and secretive terrorist cells that can ignore the need for mass mobilisation in its politics of terror, or it is the method of relatively isolated groups such as the adivasis of central India, who may create liberated zones in their forest and mountain tracts, but have little or no capability of extending such a politics into the wider society beyond. Non-violence also encourages dialogue and negotiation, and does not alienate potential allies. It is, thus, a far more effective force for building a future democracy.

The celebratory tone of much of the early SSC work towards violence then has needed to be redressed. This is important for the later context of this book, where the propagation of violent rebellion was put forward by exactly the groups of idealistic young men identified by Hardiman. The four main individuals who shaped the geographies of the Pondicherry Gang here were all able-bodied men at the start of their revolutionary careers, and all espoused at least some form of violent, and frequently masculinised, resistance to colonialism. However, their encounters with and responses to (anti)colonial violence were all different – and some were broken physically and mentally by the violent colonial response to their agitation. What the politics of the group exposes then is that, whilst the anticolonialism of the Pondicherry 'Gang' was productive of new and dynamic visions of the world, it also illustrates some of the very real limits of the masculinised violence which the group deployed. This will become clear throughout the later chapters of this book.

However, despite these important critiques, the broad sweep of its anti-elitist approach meant that the SSC has had a huge impact upon politics, history and many related areas – as noted in the previous chapter, the Latin American Subaltern Studies Group was at least partially inspired by the original SSC. Geography has been relatively slow to engage in real depth with subaltern ideas, despite the usage of the term subaltern being common in geography for some time. The exceptions here are the likes of Clayton (2011), whose work has long interrogated issues of subalternity and geographical knowledge production. Elsewhere, David Featherstone (2009, 2012, 2017) has pioneered work on subaltern forms of cosmopolitanism – tracing transnational networks of labourers and workers across space. Subaltern forms of cosmopolitanism have also been a focus of Gidwani (2006, 2009), and this work alongside Featherstone's has usefully challenged the idea that cosmopolitanism is only ever an elite practise. The postcolonial impact on thinking the urban differently has meant some urban geographers have been more attentive to these discussions (Robinson 2003, 2016). Tariq Jazeel (2014) has also used the subaltern to call for a greater attentiveness to how translations of many of the key terms of geography often obscure much of the nuance and specifics that emerge from particular cultural contexts, in his case in understandings of 'nature' in Sinhalese Sri Lanka. Joanne Sharp's work on subaltern and feminist geopolitics (Sharp 2011, 2013) has used ideas from the global south to write back and challenge conventional and masculinist geopolitics. Most recently, and emerging from sessions at the RGS–IBG Conference, Jazeel and Legg (2019) have edited, and the first collection of geographical work specifically focussed on the subaltern.

Subaltern geography then is still unfolding as a grouping of analyses. Whilst this book does not necessarily identify as a part of this literature – many of the individuals in its pages would be 'elite' in many ways – it does articulate with this growing body of work. Not the least of this is how the reading of anticolonialism in the last chapter is indebted to a subaltern political ethics. I approach this work similarly to Chari (2011) and Featherstone (2019) in claiming that it is important to push at how subaltern practises can actively reshape space and society – thus, that subaltern practises, although minor, articulate and create new political spaces. Thus, to me, subaltern geography should not remain wedded to the idea that subalternity was an 'autonomous domain', cut off from other, hierarchically superior, social groupings. This, as Featherstone (2019) points out means that the figure of the subaltern becomes siloed off and unable to engage politically. Instead, we should recognise that subaltern politics takes place – both in that it occurs, but that it also attempts to reconfigure, in situated ways, the world that has created conditions of subalternity.

The subaltern then is of twofold importance to this book. Firstly, the intention of writing histories from below which were at the heart of the initial impetus of the SSC opened up the social sciences as a whole, not just history, which has fundamentally impacted the contents of this book. As will become clear in the next section on Tamilian/South Indian historiography, the impact of the SSC has changed the nature of how anticolonial and other radical political forms operated in South India. However, in a more geographical sense, the book is indebted to the subaltern approaches that have sought to open up space for subaltern knowledges and peoples in geographical scholarship, but seeks to carry on the expansion of them through an engagement with the political. Whilst to a large extent, the individuals discussed in the empirical chapters are bourgeois–elitist – the vast majority of them were writing newspapers and performing formal politics in an attempt to mobilise people to take part in anticolonial politics before the Gandhian phase of the freedom struggle began true mass participation – their position as forgotten yet still important figures who often failed in their various enterprises are important, subaltern, aspects of the struggle for Indian independence. In order to provide some wider context for these empirical chapters, the final section of this chapter looks more specifically at the South Indian context to these men's lives.

Tamilian South India – Politics and Society in a 'Backwater' of Historical and Geographical Research

Madras Presidency was the British territorial jurisdiction that governed most of Southern India during the majority of the colonial era (see Map 3.1). Governed from Fort St George in Madras, the Presidency was ruled by a state-appointed Governor, who was supported by a legislative council who were collectively responsible for maintaining day to day British rule, including policing, trade and infrastructure. Bordering and enclaved within the Presidency, there were also a number of Princely States, ruled by 'Native' Princes or Maharajas who had sworn allegiance to the British crown which were adjacent to the Presidency, most notably Hyderabad (which became part of the present states of Andhra Pradesh and Telangana), and Banganapalle in the north, Mysore, (part of present-day

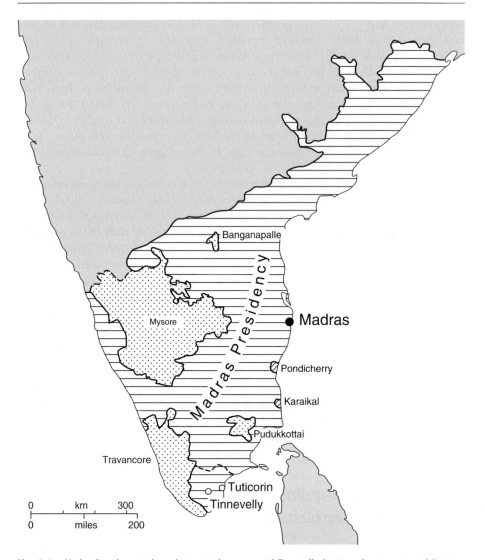

Map 3.1 Madras Presidency in the early twentieth century, with Tinnevelly district and major territorial divisions noted. Source: Map prepared by Suzanne Yee.

Karnataka), Travancore in the SouthWest (in present day Kerala), and Pudukkottai the south (in present-day Tamil Nadu) (see Dirks 1987 for a discussion of politics in these Princely States). There were also foreign enclaves within the Presidency in the form of three French *comptoirs* (the small enclaves of Pondicherry and Karaikal on the Eastern Coast, and the even tinier town of Mahe on the west) which will be discussed in the next chapter. At various other times, Dutch,

Portuguese and Danish enclaves also existed, although these had been subsumed into British India by the turn of the twentieth century.

The Presidency itself was diverse and reflected the gradual expansion of British power during the eighteenth and nineteenth centuries, so that there were a variety of different ethno-linguistic and social groupings within the state. These were reflective of the distinctive 'Dravidian' socio-linguistic structure of South India. The four major languages of the south of India – Tamil, Telugu, Malayalam and Kannada – are collectively grouped as Dravidian languages. This classification emerges from orientalist histories of India which adopted the mythic past of an Aryan migration into (or sometimes invasion of) India from Central Asia during pre-history (Sanyal 2012). This migration supposedly forced the indigenous 'Dravidian' inhabitants of India southwards. Seizing upon this mythical past, British colonial administrators like Herbert Risley used racial science to divide Dravidian and Aryan peoples and much of this narrative of Dravidians as a somehow 'older' population than those present in North India still circulates in popular culture and is utilised by populist Dravidian politicians today.

The majority of this book explores the lives of a number of individuals who were from Tamil-speaking areas of the Presidency, and which broadly conform to the area of the present state of Tamil Nadu. In the intervening 100 years or so, Tamil identity has morphed and altered dramatically, with political shifts of power attempting to challenge orthodox Tamil Brahmin social practises, and post-independence, a rise in Tamil nationalism against the perceived 'Hindi imperialism' of the northern states of India has hardened a sense of Tamil, or at least wider Dravidian, exceptionalism for many in the South. However, Tamil identity is tied into the complex and distinctive language structures which are traced back to the Sangam poetry era (c. 300 BCE–300 CE), and although there are debates about how Sanskrit and Tamil share certain words and terms, Tamil is still seen as classical language in its own right (Shulman 2016). Again, Tamil distinctiveness also functions through the use of historical dynasties like the medieval Chola Kingdom/Empire (c. 848–1279 CE) based in the Kaveri river delta and which ruled large parts of Southern India and even extended its influence to South East Asia which is still often marked out as an important political and trading polity that belong to this Tamil/Dravidian historical past.

Post-independence, and outside the purview of this book, but signalling how this identity has become politicised, Tamil political parties like the Dravida Munnetra Kazhagam (Dravidian Progressive Conference, or DMK), and later it's splinter and arch-competitor the All India Anna Dravida Munnetra Kazhagam (AIADMK) formed the vanguard of a Tamil 'nationalist' movement, mobilised through resistance to the central Government of the Republic of India's attempts to promote Hindi as the sole national language (Guha 2007). Tamil identity then is complex and often exists in an uneasy relationship with modern pan-Indian (and Hindi-dominated) identities that (to Tamils) often denigrate this cultural and political history.

Tamil-speaking areas of India also have their own distinctive caste structure. The typical caste structure of India is often imagined through the orientalist and problematic construction of the caste system as a series of '*Varnas*' (*Brahmins* or the priestly caste, *Kshatriyas* or warriors, *Vaishyas* or farmers/traders, and *Shudras* or labourers) each of which are made of thousands of *Jatis* or distinctive sub-groups. Existing outside of this are the Scheduled Castes or *Dalit-Bahujans* – the current term meaning 'broken' (Dalit) and 'the majority of society' (Bahujan), for what in the colonial era were termed 'untouchable' communities who were and are subject to intense forms of social discrimination. The *varna* categorisation is now largely seen as a problematic orientalist construct (Jodhka 2012), yet still has important legacies in determining the reservation of jobs for members of *Dalit-Bahujan* and other marginalised or subaltern communities.

In Tamil-speaking areas, the usefulness of the fourfold structure of the *Varnas* breaks down. There are notably few, if any, representatives of the *Kshatriya varna*. Instead, *Shudras* occupy a large proportion of positions of socio-economic power. Brahmins, especially in the period under study, were still at the top of the hierarchy. Tamil Brahmins (often abbreviated to 'Tam Brams') were notably orthodox in their interpretations of the rules of religious and cultural life. The nineteenth century saw migration of Brahmin groups from rural locations to cities (Fuller and Narasimhan 2014), and the ensuing stratification of them into positions of financial and professional power meant that any social interaction between Brahmins and other, especially *Dalit-Bahujan*, groups was minimised. The resultant social stratification proved fertile ground for the non-Brahmin and anti-caste movement that emerged in the state, such as that galvanised by the middle-class firebrand E.V. Ramaswamy, more often known by the epithet 'Periyar'. Combined with this, there was also considerable fragmentation of society based on patterns of traditional agriculture and access to water, with river valleys producing different social structures to the more climatically unreliable plains and hills (see the discussion of caste and population patterns in Gorringe 2017).

Lastly, shifts within indigenous industry also shifted under colonial rule. Tamil industry was dominated by a highly skilled handloom and weaving industry which was severely disrupted by the industrial revolution taking place in the United Kingdom (Beckert 2014). In the latter half of the nineteenth century, domestic markets for these industries collapsed as cheap British-produced goods from Lancashire mills swamped the market (Ganeshram 2017). This created huge shifts in the labour force in the Presidency, with concomitant swathes of impoverishment and destitution. All of this meant that by the early twentieth century, once ideas about the supposed 'drain' (Chandra 1965) of India by the British were being mobilised by Swadeshi activists, this collapse of 'indigenous' industry was within living memory and was a ready repository of anti-British grievance, which will become clear in Chapter Four.

To turn more closely to political activity, this final part of the chapter also challenges the continuing absence of studies that examine the South of India during the freedom struggle. To an extent, this is to be expected given the seeming lack of 'key' moments of nationalist agitation in the South. Various figureheads came from the south (e.g. C. Rajagopalachari, the Governor-General of India when it gained independence in 1947, was born in Madras Presidency), and there were moments of mass mobilisation at various times, some of which will be discussed during the course of this book. But, compared to the wave of 'revolutionary terrorism' (Heehs 1992, 1994) in Bengal following its partition in 1905, or the mass *Satyagraha* movements after Mohandas Gandhi returned to India from South Africa in 1915, the South was typified less by mass movements and more by social and political reform movements, such as the anti-Brahmin Justice Party, which formed in 1917 and campaigned for a number of social reforms and was opposed to traditional caste hierarchies (Pandian 2007). The impact of the Justice Party, particularly as it spawned a distinctly Tamil political milieu with the formation of the DMK and has shaped subaltern and mainstream politics in the region and across India, means that this history tends to have attracted most of the academic scholarship of the region, at the expense of, for example, Dalit histories (Gorringe 2017). Indeed, the later chapters of this book can be read to some extent as four examples of failure in anticolonial mobilisation – only two of the men discussed survived to see Indian independence, and by that time, one (M.P.T. Acharya) was living in poverty having seen his dreams of international anarcho-socialist revolution lead nowhere, whilst the other (Aurobindo Ghose) was living as a spiritual guru. As a result, the pre-Gandhian wave of anticolonialism in South India tends to be seen as a brief and shortlived experiment before the vast majority of agitations moved on to elsewhere in colonial India. It is not surprising that Subramania Bharati would describe the pre-Gandhian anticolonial movement in the South as 'stillborn' (cited in Venkatachalapathy 2010, p. 37).

However, to dismiss the development of anticolonialism (and its corollary, nationalism) in South India as marginal also occludes the dynamic nature of political activity taking place during the fin-de-siècle era, and how this 'political' activity was shaped across social, cultural and political avenues. There has, of course, been work done on political formations in the south of India previously. However, much of this has not been explicitly 'anticolonial', but has instead been a study of administrative histories, or of the emergence of the particular counterhegemonic (to both colonial and Hindi) Dravidian political movements that began around 1920 in South India. The most notable critique of the former of these tendencies lies in the so-called Cambridge School's writings, particularly David Washbrook's and Christopher Baker's attempts to understand the development of political activity from mid-nineteenth century until India's independence in 1947 (Washbrook and Baker 1975; Baker 1976; Washbrook 1976). These works focussed primarily on a very limited and formal idea of politics as viewed from the administration of the colonial state, and as a result, exclude socio-political issues that existed beyond this category. Ranajit Guha (1998) strenuously critiqued Washbrook in particular for his attempts to evacuate the social from this history, which in turn removes the ideological from the history of the political in India. Similarly, M.S.S. Pandian (1995) argued that the Cambridge School's attempts to emphasise factionalism and clientelism as mechanisms of political organisation in Madras Presidency meant that they omitted the role of caste from their analysis, especially in the study of the emergence of the Justice Party and Self-Respect movements in the region. Importantly to Pandian, the key to understanding the political in South India (as elsewhere) is to avoid separating the cultural and the political into distinct spheres, and to him, scholars like Washbrook and Baker are guilty of setting up a narrowly defined version of 'the political' that does not allow for an understanding of the core interrelationship between it and other 'zones' of social behaviour. This again shows the limits to the Political framings that are applied to subaltern or colonised subjects.

As a result, Madras Presidency, and the south of India in general, deserves to be re-evaluated and seen as more than just a peripheral zone, or as somehow different or disconnected from the rest of India. Indeed, the South of India was closely interconnected with currents of anticolonial thought and practise which stretched across India and beyond. Many of the various anticolonialists who were either born in or settled in the Presidency are either important figures in their own right, or at least deserve more scholarly attention. These included individuals like Aurobindo Ghose, V.V.S. Aiyar, Subramania Bharati, V.O.C. Pillai, G. Subramania Iyer, M.P.T. Acharya, Srinivas Acharya and many more. Some of these names will become familiar during later chapters of this book, but it should be noted that there is a general absence of women within this milieu, and whilst Western women like Annie Besant of the Theosophical Society, and Mirra Alfassa, Aurobindo's consort after his arrival in Pondicherry, would become important in this South Indian context (for different reasons), there is some work to be done on recognising the role of Indian women in supporting this political movement.

However, there is also a degree of parochialism at work within work on the south of India, largely driven by the break-up of the Presidency in the postcolonial era into the states of Kerala, Tamil Nadu, Andhra Pradesh, Karnataka and, most recently, Telangana. This has created a tendency to post-hoc rationalise what anticolonial activity there has been in these geographical areas into discrete, localised or vernacular groupings. For example, Rajendran's *National Movement in Tamil Nadu* (1994) does useful work in collating in one book the range and scope of the more radical/extreme aspects of anticolonial activity in Tamil speaking areas of Madras Presidency. However, it also emphasises the 'national' as the key domain through which anticolonial resistance principally played out, and as a result, misses some of the core objections to the elite, nationalist historiography critiqued by the Subaltern Studies group. A similar spatio-political confusion emerges in Rajendran's use of Tamil Nadu. This translates as 'Tamil Land' or 'Tamil Country', but in being used this way emphasises the postcolonial political structure of the state of Tamil Nadu, which was formed in 1969 as a renaming of

Madras State, which was in turn created as part of the ongoing break-up of the larger Madras Presidency into smaller linguistically based states post-independence. Tamil nationalist identity in the period Rajendran studies (1905–1914) did not exist in the shape of a uniform set of discourses and practices, and when 'nationalism' was mobilised by those at work in the south, it was distinctly tied into pan-Indian identities, such as the Swadeshi movement. This also tends to infer that anticolonial activities taking place in Tamil Nadu were discretely 'Tamil' and somehow different or distinct from other movements across the southern peninsula of India. This, I would argue, is driven partially by a sense of Tamil exceptionalism which has emerged from both a sense of Tamil grievance against a Hindi- and Bengali-dominated postcolonial narrative of the freedom struggle and is a legacy of Tamil and Dravidian nationalist movements which have long argued that Southern India's distinctive linguistic culture has been oppressed within the context of dominant Hindi culture in the Republic of India.

This book takes a slightly different approach by spatially locating much of its narrative in Tamil-speaking areas of India, (and indeed, three of the four major figures examined in this book were born in Tamilian areas of India), but will stretch the narratives associated with these revolutionary networks across and beyond India itself. It does this for two reasons. Firstly, and somewhat self-evidently, the study of the emergence of Indian anticolonialism in the first decades of the twentieth century necessarily means that connections to the emerging idea of India as a 'national space', and thus, separating Tamil areas from the rest of India imposes a false territorial distinction between them. Secondly, and drawing on a wider turn in the literature towards more globally oriented and interconnected histories/geographies discussed previously, the book's wider argument seeks to place the anticolonialisms it discusses within wider networks, and this decentres and challenges any fixed notion of anticolonialism being somehow 'nation-centric', or in this case, 'Tamil-centric'. Instead, it looks at the anticolonial geographies which were produced through the distinctly Tamil spaces and places which are the subject of this book. Whilst not wishing to fetishise Tamil Nadu/Madras Presidency-based anticolonial movements as somehow exceptional, it is important to recognise the distinctiveness of them, lots of which, as will become clear throughout this book, is due to the particular patterns of culture and politics that existed both in the city of Madras and elsewhere in its Presidency. In this, the book builds on an extensive amount of work done by a number of scholars which it is important to recognise here, both in terms of providing a background to readers unfamiliar with the context, but also in order to prove that Madras Presidency deserves more attention in this regard.

The orthodox narrative of the emergence of Indian 'nationalism' in Tamil South India is comprised of a number of different societal and political changes which occurred alongside the broader shifts in society, culture, economy and politics which were discussed previously. Most recently, Ganeshram (2017) has examined socio-economic factors leading to the emergence of 'nationalism' in the period 1858–1918, arguing that the main arenas driving this change are changing patterns in, and the increasing Westernisation/Christianisation, of education; the increasing role of the press; socio-religious reform movements aimed at liberalising issues like caste-based injustice; the decline of indigenous industries, and; changes in colonial agrarian policy. These processes all indicate the broad changes which were occurring as 'Western modernity' (broadly conceived) intervened in and reshaped existing Tamil society. Whilst large-scale processes such as the decline of the handloom industry discussed previously had clear socio-economic consequences, much of this change was considerably more quotidian in nature. For example, Venkatachalapathy (2006) explores how the emergence of drinking coffee in the modern era, as well as altering drinking customs, marked a shift in broader cultural norms and practises across middle class Tamil society, but also acted to reinscribe coffee as a middle class consumer product, whilst tea was distinctly working class. This meant that challenges to orthodox Tamil Brahminism and its highly regimented social codes governing who was allowed to eat what, with who, and where, began to emerge alongside broader socio-economic trends. Tamil-speaking areas of India were not unique in experiencing these changes, but the distinctiveness of Tamil culture did mean that these changes played out in geographically contingent ways. Many of these changes will form part of the bedrock of later chapters.

Conclusions

This chapter has introduced a number of key aspects of recent scholarship concerning the struggle for Indian independence. I have focussed upon issues related to non/violence, inter/nationalism and the elite/subaltern, and these by no means give a full account of the variety of ways in which one could study the Indian anticolonial movement. However, these areas do provide some distinct overlaps with debates and concerns which are important to geographical scholarship, with spatial concerns at the heart of each. As such, the review sections of this chapter have served largely to introduce these as broad thematic areas of note, and many of them will be returned to or elaborated upon in the chapters that follow. In addition to this, the chapter has done some important work for the book as a whole in beginning to develop a sense of the spaces and places in which the theoretical work discussed so far is going to be discussed through. Whilst the Tamilian sections of Madras Presidency do not form the main geographical backdrop to all of the four empirically driven chapters that follow, this region forms an anchor around which the various networks and individuals included in these chapters gravitated around at some point in their lives. Thus, the final section of this chapter necessarily presented some of the key characteristics of Tamil society, both during the period of study and after. This served to both familiarise the region to those who are unaware of it, but also this performs a political task of ensuring that this often neglected or marginal space is given its due as an important hub for anticolonial ideas and practices. The next chapter begins the more detailed exploration of how these anticolonial political practises played out. In this case, we will explore how the emergence of 'nationalism' in the form of the Swadeshi Steam Navigation Company exceeded the territorial and 'landed' spaces of India.

References

Abraham, I. (2015). "Germany has become Mohammedan": insurgency, long-distance travel, and the Singapore Mutiny, 1915. *Globalizations* 12 (6): 913–927. https://doi.org/10.1080/14747731.2015.1100850.

Ahmed, R. and Mukherjee, S. (eds.) (2012). *South Asian Resistances in Britain, 1858–1947*. London: Continuum.

Allen, C. (2018). Who owns India's history? A critique of Shashi Tharoor's inglorious empire. *Asian Affairs* 49 (3): 355–369. https://doi.org/10.1080/03068374.2018.1487685.

Amin, S. (1995). *Event, Metaphor, Memory: Chauri Chaura, 1922–1992*. University of California Press.

Anderson, C. (2012). *Subaltern Lives: Biographies of Colonialism in the Indian Ocean World, 1790–1920*. Cambridge: Cambridge University Press.

Anim-Addo, A. (2014). "The great event of the fortnight": steamship rhythms and colonial communication. *Mobilities* 9 (3): 369–383. https://doi.org/10.1080/17450101.2014.946768.

Arnold, D. (1986). *Police Power and Colonial Rule: Madras 1859–1947*. Bombay: Oxford University Press.

Arnold, D. (1994). The colonial prison: power, knowledge and penology in nineteenth century India. In: *Subaltern Studies VIII: Essays in Honour of Ranajit Guha* (eds. D. Arnold and D. Hardiman), 148–187. Oxford: Oxford University Press.

Baker, C.J. (1976). *The Politics of South India: 1920–1937*. Cambridge: Cambridge University Press.

Bate, B. (2012). Swadeshi in the Time of Nations. *Economic and Political Weekly* 47 (42): 42–43.

Bayly, C.A. (2004). *The Birth of the Modern World*. Oxford: Blackwell.

Beckert, S. (2014). *Empire of Cotton: A Global History*. New York, NY: Knopf.

Bose, S. and Manjapra, K. (eds.) (2010). *Cosmopolitan Thought Zones: South Asia and the Global Circulation of Ideas*. Basingstoke: Palgrave Macmillan.

Brown, J. (1977). *Gandhi and Civil Disobedience: The Mahatma in Politics 1928–34*. Cambridge: Cambridge University Press.

Brückenhaus, D. (2017). *Policing Transnational Protest: Liberal Imperialism and the Surveillance of Anticolonialists in Europe, 1905–1945*. Oxford: Oxford University Press.

Burton, A. and Ballantyne, T. (eds.) (2016). *World Histories from Below: Disruption and Dissent, 1750 to the Present*. London: Bloomsbury.

Chakrabarty, D. (2007). *Provincialising Europe: Postcolonial Thought and Historical Difference*, 2e. Princeton, NJ: Princeton University Press.

Chandra, B. (1965). Indian nationalists and the drain, 1880–1905. *Indian Economic and Social History Review* 2 (2): 103–144.

Chandra, B. (2012). *The Writings of Bipan Chandra: The Making of Modern India: From Marx to Gandhi*. Hyderabad: Orient Blackswan.

Chandra, B. et al. (1989). *India's Struggle for Independence*. New Delhi: Penguin.

Chari, S. (2011). Subalternities that matter in a time of crisis. In: *The New Companion to Economic Geography* (eds. J. Peck, T. Barnes and E. Sheppard), 501–514. Chichester: Wiley-Blackwell.

Chatterjee, P. (1993). *The Nation and Its Fragments*. Princeton: Princeton University Press.

Chaturvedi, V. (2000). *Mapping Subaltern Studies and the Postcolonial*. London: Verso.

Chaturvedi, V. (2013). A revolutionary's biography: the case of V D Savarkar. *Postcolonial Studies* 16 (2): 124–139. https://doi.org/10.1080/13688790.2013.823257.

Chibber, V. (2013). *Postcolonial Theory and the Spectre of Capital*. London: Verso.

Clayton, D. (2011). Subaltern space. In: *The Sage Handbook of Geographical Knowledge* (eds. J. Agnew and D. Livingstone), 246–260. London: Sage.

Davies, A.D. (2012). Assemblage and social movements: Tibet Support Groups and the spatialities of political organisation. *Transactions of the Institute of British Geographers* 37 (2): 273–286. https://doi.org/10.1111/j.1475-5661.2011.00462.x.

Davies, A.D. (2013). Identity and the assemblages of protest: the spatial politics of the Royal Indian Navy Mutiny, 1946. *Geoforum* 48: 24–32. https://doi.org/10.1016/J.GEOFORUM.2013.03.013.

Davies, A. (2019). Transnational connections and anti-colonial radicalism in the Royal Indian Navy mutiny, 1946. *Global Networks*.

Dirks, N.B. (1987). *The Hollow Crown: Ethnohistory of an Indian Kingdom*. Cambridge: Cambridge University Press.

Elam, J.D. and Moffat, C. (2016). On the form, politics and effects of writing revolution. *South Asia: Journal of South Asian Studies* 39 (3): 513–524. https://doi.org/10.1080/00856401.2016.1199293.

Featherstone, D.J. (2008). *Resistance, Space and Political Identities: The Making of Counter Global Networks*. Oxford: Wiley-Blackwell.

Featherstone, D.J. (2009). Counter-Insurgency, subalternity and spatial relations: interrogating court-martial narratives of the Nore mutiny of 1797. *South African Historical Journal* 61 (4): 766–787.

Featherstone, D.J. (2012). *Solidarity: Hidden Histories and Geographies of Internationalism*. London: Zed Books.

Featherstone, D. (2015). Maritime labour and subaltern geographies of internationalism: black internationalist seafarers' organising in the interwar period. *Political Geography* 49: 7–16.

Featherstone, D. (2017). Anti-colonialism and the contested spaces of communist internationalism. *Socialist History* 52: 48–58.

Featherstone, D. (2019). Reading subaltern studies politically: histories from below, spatial relations, and subalternity. In: *Subaltern Geographies* (eds. T. Jazeel and S. Legg), 94–118. Athens, GA: University of Georgia Press.

Fischer-Tine, H. (2007). Indian nationalism and the "world forces": transnational and diasporic dimensions of the Indian freedom movement on the eve of the First World War. *Journal of Global History* 2 (3): 325–344.

Fuller, C.J. and Narasimhan, H. (2014). *Tamil Brahmins: The Making of a Middle-class Caste*. Chicago: University of Chicago Press.

Gandhi, L. (2006). *Affective Communities: Anticolonial Thought, Fin-de-Siecle Radicalism, and the Politics of Friendship*. London: Duke University Press.

Ganeshram, S. (2017). *Pathways to Nationalism: Social Transformation and Nationalist Consciousness in Colonial Tamil Nadu, 1858–1918*. Oxon: Routledge.

Ghosh, D. (2016). Gandhi and the terrorists: revolutionary challenges from Bengal and engagements with non-violent political protest. *South Asia: Journal of South Asian Studies* 39 (3): 560–576. https://doi.org/10.1080/00856401.2016.1194251.

Ghosh, D. and Kennedy, D. (eds.) (2006). *Decentring Empire: Britain, India and the Transcolonial World*. London: Sangam Books.

Gidwani, V. (2006). What's left? Subaltern cosmopolitanism as politics. *Antipode* 38 (1): 8–21. https://doi.org/10.1111/j.0066-4812.2006.00562.x.

Gidwani, V. (2009). Subalternity. In: *International Encyclopedia of Human Geography* (eds. R. Kitchin and N. Thrift), 65–71. Oxford: Elsevier.

Gorringe, H. (2017). *Panthers in Parliament: Dalits, Caste and Political Power in South India*. Oxford: Oxford University Press.

Goswami, M. (2004). *Producing India: From Colonial Economy to National Space*. Chicago: University of Chicago Press.

Goswami, M. (2012). Imaginary futures and colonial internationalisms. *The American Historical Review* 117 (5): 1461–1485. https://doi.org/10.1093/ahr/117.5.1461.

Guha, R. (1992). Discipline and mobilise. In: *Subaltern Studies VII* (eds. P. Chatterjee and G. Pandey), 69–120. New Delhi: Oxford University Press.

Guha, R. (1998). *Dominance without Hegemony: History and Power in Colonial India*. Delhi: Oxford University Press.

Guha, R. (2007). *India after Gandhi*. London: Macmillan.

Hardiman, D. (2013). Towards a history of non-violent resistance. *Economic and Political Weekly* XLVIII (23): 41–48.

Heehs, P. (1992). The Maniktala secret society: an early Bengali terrorist group. *The Indian Economic and Social History Review* 29 (3): 349–370.

Heehs, P. (1994). Foreign influences on Bengali revolutionary terrorism 1902–1908. *Modern Asian Studies* 28 (3): 533–556. https://doi.org/10.1017/S0026749X00011859.

Hodder, J. (2016). Toward a geography of black internationalism: Bayard Rustin, nonviolence, and the promise of Africa. *Annals of the American Association of Geographers* 106 (6): 1360–1377. https://doi.org/10.1080/24694452.2016.1203284.

Iton, R. (2008). *In Search of the Black Fantastic: Politics and Popular Culture in the Post-Civil Rights Era*. Oxford: Oxford University Press.

Jazeel, T. (2014). Subaltern geographies: geographical knowledge and postcolonial strategy. *Singapore Journal of Tropical Geography* 35 (1): 88–102. doi:10.1111/sjtg.12053.

Jazeel, T. and Legg, S. (eds.) (2019). *Subaltern Geographies*. Athens, GA: University of Georgia Press.

Jodhka, S.S. (2012). *Caste*. New Delhi: Oxford University Press.

Kachwala, S. (2018). Recovering history: gender, anti-colonial militancy and Indian popular cinema. *Gender & History* 30 (3): 704–717. https://doi.org/10.1111/1468-0424.12400.

Kaiwar, V. (2015). *The Postcolonial Orient: The Politics of Difference and the Project of Provincialising Europe*. Chicago: Haymarket.

Kapila, S. (2010). A history of violence. *Modern Intellectual History* 7 (2): 437–457. https://doi.org/10.1017/S1479244310000156.

Kaviraj, S. (1992). The imaginary institution of India. In: *Subaltern Studies VII* (eds. P. Chatterjee and G. Pandey), 1–39. Oxford: Oxford University Press.

Kothari, U. (2011). Contesting colonial rule: politics of exile in the Indian Ocean. *Geoforum* 43 (4): 697–706. https://doi.org/10.1016/j.geoforum.2011.07.012.

Legg, S. (2007). *Spaces of Colonialism: Delhi's Urban Governmentalities*. Oxford: Wiley-Blackwell.

Legg, S. (2009). Of scales, networks and assemblages: the League of Nations apparatus and the scalar sovereignty of the Government of India. *Transactions of the Institute of British Geographers* 34 (2): 234–253. https://doi.org/10.1111/j.1475-5661.2009.00338.x.

Legg, S. (2014). An international anomaly? Sovereignty, the League of Nations and India's princely geographies. *Journal of Historical Geography* 43: 96–110. https://doi.org/10.1016/j.jhg.2013.03.002.

Lester, A. (2001). *Imperial Networks: Creating Identities in Nineteenth Century South Africa and Britain*. London: Routledge.

Lester, A. (2005). Imperial circuits and networks: geographies of the British Empire 1. *History Compass* 4 (1): 124–141. https://doi.org/10.1111/j.1478-0542.2005.00189.x.

Maclean, K. (2015). *A Revolutionary History of Interwar India:Violence, Image,Voice and Text*. London: Hurst & Company.

Maclean, K. (2016). Revolution and revelation, or, when is history too soon? *South Asia: Journal of South Asian Studies* 39 (3): 678–694. https://doi.org/10.1080/00856401.2016.1191536.

Maclean, K. and Elam, J.D. (2013). Reading revolutionaries: texts, acts, and afterlives of political action in late colonial South Asia. *Postcolonial Studies* 16 (2): 113–123. https://doi.org/10.1080/13688790.2013.823259.

Mahajan, G. (2013). *India: Political Ideas and the Making of a Democratic Discourse*. London: Zed Books.

Mohanty, S. (2015). *Cosmopolitan Modernity in Early 20th Century India*. Oxford: Routledge.

Mukherjee, S. (2018). *Indian Suffragettes: Female Identities and Transnational Networks*. Oxford: Oxford University Press.

O'Hanlon, R. (1985). *Caste, Conflict and Ideology: Mahatma Jotirao Phule and Low Caste Protest in Nineteenth Century Western India*. Cambridge: Cambridge University Press.

O'Malley, K. (2008). *Ireland, India and Empire: Indo-Irish Radical Connections, 1919–1964*. Manchester: Manchester University Press.

Osterhammel, J. (2014). *The Transformation of the World: A Global history of the Nineteenth Century*. Princeton, NJ: Princeton University Press.

Pandian, M.S.S. (1995). Beyond colonial crumbs: Cambridge school, identity politics and Dravidian Movement(s). *Economic and Political Weekly* 30 (7/8): 385–391.

Pandian, M.S.S. (2007). *Brahmin and Non-Brahmin: Genealogies of the Tamil Political Present*. Ranikhet: Permanent Black.

Petersson, F. (2014). Hub of the anti-imperialist movement. *Interventions* 16 (1): 49–71. https://doi.org/10.1080/1369801X.2013.776222.

Prashad, V. (2008). *The Darker Nations: A People's History of the Third World*. New York, NY: The New Press.

Rajendran, N. (1994). *National Movement in Tamil Nadu, 1905–1914: Agitational Politics and State Coercion*. Madras: Oxford University Press.

Ramaswamy, S. (2010). *The Goddess and the Nation: Mapping Mother India*. Durham, NC: Duke University Press.

Ramnath, M. (2011a). *Decolonizing Anarchism: An Antiauthoritarian History of India's Liberation Struggle, Anarchist Interventions*. Edited by I. for A. Studies. Edinburgh: AK Press.

Ramnath, M. (2011b). *Haj to Utopia: How the Ghadar Movement Charted Global Radicalism and Attempted to Overthrow the British Empire*. Berkeley: University of California Press.

Raza, A., Roy, F., and Zacharia, B. (2015). Introduction. In: *The Internationalist Moment* (eds. A. Raza, F. Roy and B. Zachariah), xi–xli. New Delhi: SAGE Publications India.

Robinson, J. (2003). Postcolonialising geography: tactics and pitfalls. *Singapore Journal of Tropical Geography* 24 (3): 273–289. https://doi.org/10.1111/1467-9493.00159.

Robinson, J. (2016). Thinking cities through elsewhere. *Progress in Human Geography* 40 (1): 3–29. https://doi.org/10.1177/0309132515598025.

Roy, T. (2018). Inglorious empire: what the British did to India. *Cambridge Review of International Affairs* 31 (1): 134–138. https://doi.org/10.1080/09557571.2018.1439321.

Sanyal, S. (2012). *Land of the Seven Rivers: A Brief History of India's Geography*. London: Penguin.

Savarkar, V.D. (1909). *The Indian War of Independence*. Bombay: Sethani Kampani.

Sharp, J. (2011). A subaltern critical geopolitics of the war on terror: postcolonial security in Tanzania. *Geoforum* 42 (3): 297–305. https://doi.org/10.1016/j.geoforum.2011.04.005.

Sharp, J.P. (2013). Geopolitics at the margins? Reconsidering genealogies of critical geopolitics. *Political Geography* 37: 20–29. https://doi.org/10.1016/j.polgeo.2013.04.006.

Shulman, D. (2016). *Tamil: A Biography*. Cambridge, MA: Harvard University Press.

Spivak, G.C. (1988). Can the subaltern speak? In: *Marxism and the Interpretation of Culture* (eds. C. Nelson and L. Grossberg), 271–313. Urbana: University of Indiana Press.

Stoler, A.L. (2008). *Along the Archival Grain: Epistemic Anxieties and Colonial Common Sense*. Princeton: Princeton University Press.

Thapar, R., Noorani, A.G., and Menon, S. (2016). *On Nationalism*. New Delhi: Aleph.

Tharoor, S. (2017). *Inglorious Empire: What the British did to India*. London: Hurst & Co.

Tickell, A. (2011). Scholarship terrorists: The India house hostel and the "student problem" in Edwardian London. In: *South Asian Resistances in Britain, 1858–1947* (eds. S. Mukherjee and R. Ahmed), 3–18. London: Continuum.

Venkatachalapathy, A.R. (2006). *In Those Days There was No Coffee: Writings in Cultural History*. New Delhi: Yoda Press.

Venkatachalapathy, A.R. (2010). In search of Ashe. *Economic and Political Weekly* 45 (2): 37–44.

Wagner, K. (2017). *The Skull of Alum Bheg*. Oxford: Oxford University Press.

Washbrook, D. (1976). *The Emergence of Provincial Politics: Madras Presidency 1870–1920*. New Delhi: Vikhas Publishing House.

Washbrook, D.A. and Baker, C.J. (eds.) (1975). *South India: Political Institutions and Political Change, 1880–1940*. Delhi: Macmillan.

Wilson, J. (2016). *India Conquered: Britains Raj and the Chaos of Empire*. London: Simon and Schuster.

Wolfers, A. (2016). Born like Krishna in the prison-house: revolutionary asceticism in the political ashram of Aurobindo Ghose. *South Asia: Journal of South Asian Studies* 39 (3): 525–545. https://doi.org/10.1080/00856401.2016.1199253.

Zachariah, B. (2015). Internationalisms in the interwar years: the travelling of ideas. In: *The Internationalist Moment* (eds. A. Raza, F. Roy and B. Zachariah), 1–21. New Delhi: SAGE Publications India.

Chapter Four
Appropriating Modernity and Development to Contest Colonialism: The *Swadeshi* Movement in South India and the *Swadeshi* Steam Navigation Company

Introduction

On the 30 March 1908 in the *Bande Mataram,* the pro-independence newspaper he edited in Bengal, Aurobindo Ghose, one of the leading members of the extremist faction of the Indian National Congress (INC) published a piece titled 'The Struggle in Madras'. In it, he argued:

> The carriage [of goods] by land cannot come into our hands without a political revolution, but if we hold the waterways, we shall not only hold an important part of the system of communications but be able to use our possession of it as a weapon against British trade if the railway is utilized against us. The instinct of the country had seized on this truth and the organization of *Swadeshi* steam services has been one of the first and most successful outcomes of the new movement. The Chittagong Company and Tuticorin Company have both been a phenomenal success and, owing to the spirit of self-sacrificing patriotism which has awakened in the hearts of the people, they have been able to beat their British rivals without entering into a war of rates, for the British steamers charging extravagantly low rates have been unable to command as much custom as the dearer *Swadeshi* services. A network of Companies holding the water carriage from Rangoon to Karachi and the Persian Gulf would soon have come into existence and the waterways of East Bengal would have been covered with boats plying from town to town in the ownership of *Swadeshi* concerns. If the *Swadeshi* Steam Navigation Company is crushed, this fair prospect will be ruined and all hope of commercial independence disappear for ever. (Ghose 2002, pp. 982–983 emphasis added)

Geographies of Anticolonialism: Political Networks Across and Beyond South India, c. 1900–1930, First Edition. Andrew Davies.

Aurobindo had been inspired to write by events earlier in March 1908. Tinnevelly (now named Tirunelveli), the furthest south of Madras Presidency's districts (see map in the previous chapter), had been a centre of *swadeshi* activism in 1907 and 1908, the largest enterprise of which was the establishment of the *Swadeshi* Steam Navigation Company (SSNCo) – the 'Tuticorin Company' which Aurobindo was referring to. Tuticorin (now Thoothukudi) was an industrial port town which also acted as a transit point between South India and Ceylon. Established in October 1906, the SSNCo was created by Valliappan Olaganathan Chidambaram (V.O.C.) Pillai as a *swadeshi* enterprise for industrial development, but which had the added potential of disrupting the British monopoly on transit on this route which was held by the British India Steam Navigation Company (BISNCo) – at the time, one of the largest shipping lines in the world. The existence of the SSNCo sparked a wave of industrial agitation and protest in Tinnevelly. When, Pillai and his fellow *swadeshi* organiser Subramania Siva were arrested on the 12 March in the district capital of Tinnevelly, it sparked a major protest in the town. Municipal buildings were burned, and the crowds threw stones at the police. Fearing that the situation was getting out of control, L.M. Wynch, the Collector and District Magistrate of Tinnevelly District, ordered the police to fire on the crowd, killing four people and injuring more. Further disturbances followed in Tuticorin in the following days. Robert Ashe, the Joint-Magistrate was involved in suppressing the protests there, including giving an order to fire on the crowd, although any casualties are less clear here, and it seems likely that he only fired warning shots at the crowd. This was to have significant consequences, which I discuss in the next chapter. Taken together, the events of February and March 1908 were the largest mass disturbances to take place in the pre-Gandhian phase of the Indian anticolonial movement in Madras Presidency.

The events of 1908 meant that the authorities were suspicious of *swadeshi*/seditious tendencies in the district for years afterwards, for example, in discussions about the cession of Chandernagore in 1913, it was noted by the Government of Bengal that Madras Presidency still saw Tinnevelly as a potential 'danger spot' (Home Political, Branch A, December 1913, Nos. 15–16). What is of interest here though is the fact that this series of events was spurred through a maritime enterprise. This forces us to think about anticolonialism (and *swadeshi* nationalism) in less terra-centric ways. The ability of a small maritime business to act as the motor for the creation of nationalist sentiments that destabilised wider districts, and arguably the whole of Madras Presidency, shows the political and cultural value attached to steamships in the early twentieth century, especially as markers of modernity and development.

In addition, the turn in human geography over the last decade towards taking the maritime and oceanic seriously is important to understand here. Whilst some of this has looked at the opportunities for inter/transnational organising and solidarity which steamship travel opened up for anticolonialists, there has been less written about how 'nationalist' ideas were inherently stretched beyond the territorial limits of the landed 'ocean' space. As Aurobindo's editorial shows, the dreams of industrial nationalist development that spurred much *swadeshi* organising were also often *inter*nationalist in nature – the future developmental dreams of the likes of Aurobindo were also looking outwards towards a regional, Indian Ocean, geopolitical economy. Anticolonialists were well aware that they lived within a wider imperial/colonial system, and thus, whilst 'nationalist' in character, spreading anticolonialist contestation beyond the territorial boundaries of the emergent boundaries of 'the nation' was clearly a part of the anticolonial geographies that revolutionaries were attempting to create.

This chapter then explores the SSNCo in relation to these wider ideas about maritime geographies/anticolonialisms and debates about *swadeshi* activism as a distinct set of political practices. Doing this, the chapter expands the range and scope of 'nationalist' anticolonial geographies, taking them off the land and out to sea. To do this, the chapter first examines the literature on maritime geographies more closely, before spending some time examining exactly what '*swadeshi*' forms of activism involved. An empirical section describes the events of 1907–1908 in more depth, before drawing out in more detail how exactly we can read the Tinnevelly 'riot' as a maritime-inflected anticolonialism.

Maritime and Indian Ocean Geographies

There has, over the last decade, been a rich series of engagements with maritime spaces. As Kimberley Peters (2010) argued in her landmark paper, there was a need to challenge the landed nature of much research in human geography, but also a requirement to take sea/ocean/ship spaces seriously. The tendency prior to this had been to think ocean space as seamless and friction free, or even worse, to ignore them as invisible spaces that exist beyond the view (or over the horizon) of the majority of the onshore population of the world. Yet, Steinberg (2013) has argued, reading the ocean and sea as a metaphorical space of connection occludes understanding these spaces as political and material arenas in which social practices take place. As a result there is now a large and increasing amount of interdisciplinary literature on ship, sea and ocean spaces. These have ranged from understanding disciplinary practices upon shipping lines (Ong, Minca, and Felder 2014), the domestic practices of home-making on board ships (Ryan 2006), ships and piracy as spaces of knowledge production (Hasty 2011), mobility and steamship 'rhythms' of arrival (Anim-Addo 2014), the intersections of infrastructure (in this case coal) and imperial power/mobility (Gray 2017), and, ships as spaces of social and cultural experimentation, as well as illegality, in the production of offshore pirate radio (Peters 2018), as well as much more.

In my own work on the Royal Indian Navy (RIN) Mutiny/uprising of 1946 (Davies 2013, 2014, 2019), I have argued that the maritime naval spaces of the

RIN provided a number of opportunities for anticolonial organising. Firstly, moving transnationally during World War Two exposed the sailors of the RIN to a variety of ideas about democracy, and secondly, the below-deck spaces of this ships, alongside often discriminatory policies of the RIN itself, helped to forge a pan-Indian solidarity amongst the sailors, who were often recruited from different parts of India. When this combined with the tense atmosphere of Indian and Pakistani nationalisms which occurred as part of what Chandra et al. (1989) term the post-World War Two 'upsurge' as independence approached, led to a widespread uprising that made it clear to the British authorities that the Indian armed forces could no longer be relied upon to maintain colonial rule. This work has also challenged the concept, drawn from Actor–Network Theory, of the ship as an 'immutable mobile' which shapes translocal networks through its travel between places. Instead, I argue that ships, whilst mobile, are not 'immutable' and are instead contingent and contested socio-material spaces (on this, see also Dittmer and Waterton 2018 on the assemblages of HMS Belfast).

This has also occurred at the same time as the turn towards global histories, and an increasing amount of work on the politics of mobility, has had a major impact in geography (Ogborn 2008; Osterhammel 2014). Within historical geography, work on the 'Atlantic World' (Ogborn 2005), inspired by Gilroy's (1993) *Black Atlantic* or Linebaugh and Rediker's *Many Headed Hydra* (2000), has exceeded and significantly advanced knowledge of the Atlantic 'world' as a conceptual frame to understand revolutionary and counter-revolutionary political networks which shaped the development of European modernity (Lambert 2005; Featherstone 2009, 2015).

The distinct maritime geographies of the Indian Ocean 'world', of more direct relevance to this chapter, have also become a subject of attention, from examining the different patterns of circulation that shaped distinct Indian Ocean cultural, economic and political practices (Pearson 2003; Balachandran 2006; Amrith 2015), not least the idea that Indian travel by seafaring was shaped by the culturally Hindu taboo of seeing the sea as *Kala Pani* or 'black water', by which travelling by sea meant the loss of caste status. This meant that only certain groups within South Asian society would move by sea, and shaped patterns of migration, such as the movement of large numbers of subaltern groups into the plantation economy through the indentured labour system (Brennan 1998). Most recently, Sharad Chari (2019) has attempted to use the Indian Ocean as a space to reinvigorate subalternity through its political economic pluralism – bringing the 'creolisation' of the region to the fore and embracing the differences and individual subjectivities that are produced in its spaces.

However, as Hofmeyr (2007, 2012) has pointed out, 'worlds' approaches create problematic assumptions that maritime interconnections are spatially limited to the 'ocean' with which they are associated, rather than broaching across them. A more dynamic way of understanding how maritime connections can be extended across space but also remain deeply connected with place has emerged in Island Studies, and this is the notion of the 'archipelago'. Thinking archipelagically involves challenging the notion that there is a discrete boundary where the land ends and the sea begins – as DeLoughrey (2001, p. 23) puts it:

> No island is an isolated isle and … a system of archipelagraphy—that is, a historiography that considers chains of islands in fluctuating relationship to their surrounding seas, islands and continents—provides a more appropriate metaphor for reading island cultures.

This reading of islands has been seized upon by a range of interdisciplinary scholars – for instance Vannini et al. (2009) utilise the concept to challenge readings of Canada as a continental nation, reading its islands and mainland and the effects of climate change upon them as a complex set of assemblages. Jonathan Pugh has also used the term to destabilise the binaries of land/sea, native/outsider through his work in the Caribbean, especially in the study of the cultural practice of the 'landship' in Barbados, as well as in the poetry of Derek Walcott (Pugh 2013, 2016). This work is important as it allows a sense by which distant places are impacted by and deeply connected with each other, whilst at the same time recognising the very 'placedness' that makes them different. At the same time, it pushes postcolonially at the supposed separateness of land and sea. Crucially, seeing places as interconnected, and that the movement of objects/knowledge/people across maritime spaces and over the 'border' between land and sea, allows us, in the case of this chapter, to think of an archipelago of anticolonialism being mobilised through the SSNCo.

Whilst this chapter is the most explicitly 'maritime' of this book, dealing as it does with a distinct maritime anticolonial enterprise, the range of approaches to the study of transnational contestation, both in the past and in the present, which these studies have exposed sits very closely with many of the book's key approaches. In Chapter Seven, these maritime connections will become clearer as we follow the travels of M.P.T. Acharya, but equally, the placed approach to understanding resistance in and through the Pondicherry Gang in the book as a whole owes much to seeing the intersections between various transcolonial archipelagic networks become assembled within its spaces. However, for now, it is necessary to spend some time explaining in more depth exactly what the *swadeshi* movement in India involved.

The *Swadeshi* Movement in India

So far, I have used the term '*swadeshi*' in this book and only briefly examined the specifics of the '*Swadeshi* Movement' as a distinct political moment/event. In this section, I expand on this to discuss the specifics of this 'movement' which occurred between c. 1905–1910 as a distinct wave of political activity which had important and long-lasting effects upon India's freedom struggle. Literally meaning 'of one's own country', *swadeshi* and its associated movement emerged as a result of a number of different political and social processes came together in the first decade of the twentieth century. Broadly, *swadeshi* activists saw India as politically and economically under-developed, and argued that the development of a national economy would encourage a period of national 'self-renewal', and would also be important to provide an economically viable India, post-independence. Thus, rather than Gandhian ideas of renewal through a rejection of 'modernity', much swadeshi activism in this stage was about encouraging economic development along developmentalist lines.

An exact periodisation of the *swadeshi* 'movement' is problematic as pre-existing political debates such as the 'drain theory' which impacted *swadeshism* were already in widespread circulation, and the tactics *swadeshism* established became widespread within Indian political movements and remain so until the present. Drain theory in particular is often linked to the INC Moderate Dadabhai Naoroji who used the term in 1867 to estimate how much of India's national wealth was being 'drained' out of it by colonial rule, through shifts in production, but also in the form of extraction in order to fund the wider British imperial project. Despite being classed as a moderate, and therefore not believing in revolutionary pathways towards Indian independence, by 1901, Naoroji was elaborating upon this in his *Poverty and Un-British Rule in India* (1901), and knowledge of the drain theory, whether it was technically correct or not, was widespread amongst Indian nationalists (Chandra 2012). By mobilising debates about the extractive nature of colonialism in these terms, the emerging *swadeshi* narrative about the need for India to develop economically formed a powerful framework for anticolonial debate.

Whilst *swadeshi*-style activism can be traced back to the mid-1800s, the Partition of Bengal in 1905 is often held to be the major catalyst to the intense and All India scale protests of the '*Swadeshi* Movement'. Whilst the partition was justified on the grounds of improving the administrative divide between districts of Bengal Presidency and Assam, there were plenty of overtly political grounds for dividing the territory (Sarkar 2010). The partition is now largely seen as an attempt by the government of the then Viceroy of India, Lord Curzon, to undertake a 'divide and rule' policy in the region. The upshot of the partition was an upsurge in nationalist activity. 'Moderate'-style actions such as the publication of pamphlets to educate people about the partition and the iniquities of British rule were widespread (Sanyal 2008). More 'extremist' forms of action emerged as the moderate forms of protest met with no change. Boycotts and the burning of imported goods – especially cloth – occurred across Bengal and elsewhere, and political assassination attempts began to occur (Heehs 1993). The ability to organise these varied activities was also assisted by the fact that rumours and plans for the Partition had been mooted and were public knowledge since 1903. Thus, by the time the Partition took place, there had been an extensive period of time to organise and educate people about the proposals.

The Partition of Bengal therefore intersected with the growth of *swadeshi* activism, and meant that Bengal in particular became the core region for militant anticolonialism at this time. The *Swadeshi* Movement was the first large scale challenge to British rule since the uprising of 1857. Crucially, by adopting platforms of what was then termed 'passive resistance' alongside violent measures, the movement showed the possible range and scope of anticolonial organising which was possible to both anticolonialists and the colonial authorities, and it helped to establish the tactics of non-violent direct action which became central to the independence movement. *Swadeshi* activism was, then, diverse in its impacts, and the specific approach chosen to study it often illuminates only certain aspects of this diversity. For example, given the catalytic effect of the Partition of Bengal on the *swadeshi* movement and the range of *swadeshi* activities as well as revolutionary movements that emerged in Bengal, it is not surprising that a majority of studies have focussed that region. Of those, Sumit Sarkar's *The Swadeshi Movement in Bengal, 1903–1908* (Sarkar 2010 [1973]) remains amongst the most definitive, and the sheer scale and depth of analysis of events in Bengal is outstanding. Written at around the same time as the emergence of the Subaltern Studies Collective, the book is not officially a part of that intellectual trend, but is similar in its approach to seeing the political as not just the 'elite' spaces of high politics, and treats the *swadeshi* movement in terms that we would today recognise as akin to the study of social movements. As Chakrabarty (2010) has written in a recent afterword to the book, it is a very detailed regional study, and as such is a product of the time in which the book was written, prior to the turns to the global and mobilities in the study of history and geography. Elsewhere, the book has had various critiques of it as missing out caste- and gender-based aspects of the movement that have emerged since and which Sarkar has recognised in later editions (Sarkar 2010). However, whilst a product of its time, *The Swadeshi Movement* is still definitive in its treatment of the regional/local context of Bengal, and of the wider *swadeshi* movement's strategies and tactics.

However, other approaches have also approached the movement differently. Bipan Chandra's work, alongside many others (Chandra 1965, 2012), has emphasised the important economic effects of *swadeshi* activism, particularly through the use of boycotts of 'foreign' produced goods and the promotion of indigenous '*swadeshi*' goods. The political and economic effects of *swadeshi* activism have formed a core part of Manu Goswami's work, which has extensively argued (Goswami 1998, 2002, 2004) that *swadeshi* activities began to articulate clearly for the first time the notion of a national economy that was distinctly Indian, where such a national-scale economy had not existed in the pre-colonial era. The intersections between the colonial economic system and the emergent 'Indian' or indigenous economy are important here, as in many ways, the colonial establishment shaped India's economy, in both its patterns of production, but in its very material structures – as Goswami details, the railway network both established a nation-wide set of transport links by which economic (and social/cultural/political) circulation could take place, shaping the national economy for generations to come – but also meant that the power of the colo- niser in shaping these relations was visible to everyone who used the trains or saw the station buildings as colonial establishments. Thus, because of modern economic development's centrality to the Movement's ideologies, any *swadeshi* forms of organising were built into a close and contested relationship with the spaces of the colonial economy. On the one hand, the building of a colonial national economy made the consequences of this more visible to the population of India, but on the other, it also provided a framework for *swadeshi* activists to make nationalist arguments about the unity of 'India' as an political-economic construct. Thus, as was the case for the SSNCo, often strategies and tactics of *swadeshi* political and economic activity involved the development of 'indige- nous' enterprises to replace British (or other foreign) enterprises and to allow the development of a productive economy which would work for the benefit of post-independence society.

This desire for reinvention chimes and overlaps with the desire for revivalism which was central to many nationalist tropes at the time. Whilst it is tempting to classify this as a form of Hindu-revivalism given the often primordial narratives of India's past greatness which were often deployed, this worked alongside the theory of the 'drain' to position the British as the despoilers of India. Whilst sim- plistic and disputed, the 'drain' worked strategically precisely because of its sim- plicity. In South India, where the collapse of the handloom and textile industry was a common memory (as discussed in Chapter Three), the 'drain' theory was a comparatively easy 'sell' to many people.

Whilst the political-economic aspects of *swadeshi* activism are well known, equally recognised are the cultural and symbolic aspects of the movement. As the first attempt to organise a form of mass movement, across India, the *Swadeshi* Movement of the early 1900s mobilised and established a number of repertoires which became staples of the later agitations in India such as the non-cooperation movement of 1921–1922, and even the Quit India movement of 1942. As a result, a variety of work, has emphasised how important *Swadeshi* Movement was to the material culture(s) of nationalist and anticolonial activism, which altered dress styles (such as the wearing of 'Indian', not 'Western'-style clothing), established nationalist symbols (such as India as 'Bharat Mata' or mother India) and more (Bayly 1986; Ramaswamy 2010).

Further to these 'cultural' aspects of *swadeshi* anticolonial resistance, Kris Manjapra has developed a sense in which the movement spurred the development of cosmopolitan intellectual networks. For example, Rabindranath Tagore's attempts to build an outward looking *swadeshism* that not only educated Indians in the values of Asian cultures but also engaged with intellectual trends from across the globe would, Tagore believed, create a form of 'brain irrigation' and allow the production of a future India that was global and cosmopolitan in outlook. As Manjapra puts it:

The productiveness of intellectual *swadeshi* lay in the insistence that Indians were not only recipients of foreign knowledge, but also creators and authors of knowledge and art forms that carried global significance. (Manjapra 2012, p. 54)

To Manjapra, this 'intellectual *swadeshi*' had significant long-term effects, not only in shaping the policies of various scientists in the post-independence era, but also in the development of cultural icons like the film-maker Satyajit Ray, whose impact on global cinema was, and is, probably the largest of any Indian director. There is more to say about how cosmopolitanism has been a useful marker for thinking through Indian nationalism in the later chapter on Aurobindo, but suffice to say here that *swadeshi* activism is now seen as polyglot and dynamic, and it chimes with wider political, social, cultural and intellectual movements.

The focus of this chapter then is the catalytic effect that *swadeshi* activism had upon the anticolonial movement in Madras Presidency. Despite being a much more limited venture compared to the Bengali wave of agitation, *swadeshism* acted as a key moment for those revolutionaries and activists in the presidency who wished to mobilise. As was noted in Chapter Three, Bernard Bate (2012, 2013) has shown how the public speeches of the *swadeshi* movement in Madras and elsewhere both helped to shape the repertoires of performance in Tamil politics, but these also forced the colonial authorities to develop new forms of writing in shorthand to keep up their surveillance practices.

This point of transition is immediately obvious looking through the Native Newspaper records collected by all the Presidency governments across India. These weekly reports collated all newspaper reports which were deemed to be of interest by the various Governments and included both the press written in English and in 'vernacular' forms (although in Madras Presidency at least the vernacular press was usually collated around a week later after it was translated). The information collected within these files formed a key method for the colonial governments to surveille their populations for signs of sedition, which became increasingly important in the years after 1905. Those newspapers which were establishing a reputation for irascibility, such as Subramania Bharati and M.P.T. Acharya's *India*, did often include articles which made reference to the 'drain' of India's economy by British rule, but such was the nature of the radical change embodied by the moment of Bengal's' Partition that after it, the term *swadeshi* regularly appeared in a variety of debates in the papers from the efficacy of boycotting British products, the best path to industrial development in India, or the reports of the speaking tours of various '*Swadeshi* lecturers', and which were occurring across the Presidency.

This chapter will detail some of these, but will spend the most time discussing Valliappan Olaganathan Chidambaram Pillai, most often called VOC (or Wa. Oo. Si. in Tamil). VOC came to prominence as the founder of the SSNCo, and so the chapter now turns towards the specific events of 1907–1908 in Tinnevelly district.

The *Swadeshi* Steam Navigation Company & the 'Tinnevelly Riot'

The SSNCo was established by VOC on the 16 October 1906, but the inspiration for the venture allegedly came from a speech by G. Subramania Iyer, the founder of *The Hindu*, and owner at the time of *Swadesamitran*, in August 1906 (Home Political, Branch A June 1908, No. 95, NAI (National Archives of India)). VOC was a relatively minor lawyer who had worked in Madras Presidency throughout his life. In order to raise funds for the SSNCo, he had travelled extensively around Madras Presidency collecting investment from *swadeshi*-supporting businessmen and managed to garner enough to buy two small steamers in December 1906 and January 1907, respectively. These were the SS *Gallia* and the SS *Lawoe*, and these European names were maintained rather than replaced with 'Indian' names, despite the fact that the BISNCo and other shipping lines associated with India used 'Indianised' names (More 2013). Details of these ships beyond this are sparse and are the same for the two smaller steam launches which the SSNCo subsequently purchased.

As an attempt to create an 'indigenous' shipping company, the SSNCo was a direct challenge aiming to disrupt the British monopoly on shipping lines between the port of Tuticorin in the Tinnevelly district of Madras Presidency and Colombo in Ceylon (Sri Lanka). The southerly location of the SSNCo's home meant that, as Khilnani (2016, p. 332) has put it in his introduction to Pillai for the BBC Radio 4 Series *Incarnations*, 'freedom-fighting in a place like Tuticorin meant something that it couldn't mean in riverine Calcutta: control of the seas'. This also was important as a symbolic challenge given the maritime nature of the British Empire. The establishment of the SSNCo, and its choice of target as the BISNCo's monopoly between India and Ceylon, was driven in part by VOC's knowledge of Tinnevelly/Tuticorin, but was also almost unbelievably optimistic. The BISNCo was a huge concern, containing over 100 steamships and dominating large sections of the shipping industry in the Indian Ocean region and beyond. The disparity between the two companies, combined with the broader upsurge in anticolonial activity caused by the *swadeshi* movement, meant that the SSNCo was almost immediately a source of interest. However, the tiny nature of the SSNCo, together with the nature of its demise (as we shall see), means that there is very little 'official' record-keeping that remains to assess it, and details of its day-to-day activities are scarce. As a result, the majority of sources of information on the SSNCo come from the Madras Native Newspaper Reports collected by the Criminal Investigation Department (CID) of the Government of Madras, as well as examples of newspapers and other sources like the cartoons in Subramania Bharati's *India* which have been collected by A.R. Venkatachalapathy (1994).

Given that the SSNCo was an attempt to mobilise 'indigenous' Indian capital, it took some considerable time for Pillai to raise enough funds to both buy the steamships for the line and then to get them to Tuticorin. It was not until the 17 April 1907 that the first steamship arrived in Tuticorin (Home Political,

Branch A June 1908, No. 95, NAI), and on the 25 May, Bharati published a cartoon in *India* showing crowds made up of a diverse range of India's cultures and religions welcoming the steamers (Venkatachalapathy 1994, p. 51). The effect of the SSNCo on Tuticorin was swift, with the District Magistrate writing at a point shortly after the arrival of the ships that:

> There is a regular boycott of shops selling English articles now in Tuticorin, and customers are stopped and asked what they want to buy [by *swadeshi*sts]. However, there is no complaint from shopkeepers and no violence is used. The police are insufficient in the town and the Inspector is weak and reported to be '*swadeshi*'. The Port Officer and the Customs officers also report their clerks to be '*swadeshi*'; but this is to be expected. (Home Political, Branch A June 1908, No. 95, NAI)

Over the following months, the SSNCo managed, it seems, to run fairly smoothly and at a degree of profit. The first mention of the SSNCo in the press that was noted by the authorities is in a copy of an article published in G. Subramania Iyer's *Swadesamitran* on the 2 July 1907 which argued that the 'feringhee' (foreigners) were directing their whole energy into persecuting *swadeshi* enterprises like the SSNCo (MNNR 1907–1908, Cambridge). Given the date, it is possible that this article was written by Subramania Bharati, who knew Pillai and had spoken with him in a number of *swadeshi* platforms in both Madras and Tinnevelly.

Whilst there is little in the archive available about what life was actually like on-board the ships of the SSNCo, based on records in the Tamil Nadu Archives, Rajendran (1994) has calculated that the various ships of the SSNCo conducted 214 voyages carrying 18,896 people between 1907 and 1909. The vast majority of these (114 and 12,624) fell in 1907–1908, the peak of the SSNCo's activity. According to More (2013), the *Lawoe* (which he spells as *Lavoe*) was a quite advanced ship fitted with electrical lightbulbs, and space for 44 first-class, 24 second-class, and 1300 third-class passengers and 2000 tons of goods, and would travel overnight between Tuticorin and Colombo regularly throughout the week.

There are also glimpses of what work the SSNCo was doing for the *swadeshi* cause. For example, on 2 September 1907 the English language, Indian-owned paper the *South Indian Mail*, published in Madras, pointed out that 'despite carping critics and the pseudo-patriots, this national concern continues to thrive wonderfully', and that the company was training a number of 'youths' in navigation, and that one recently trained by the Company had found employment on another steamship as a Third Officer (MNNR 1907–1908, Cambridge). It is clear from this article that the *swadeshi* goal of self-development to build a capable national economy was being employed by the SSNCo in terms of training people in order to allow them to gain employment in modern industries. That one person had been employed shows that the SSNCo's methods had a degree of respectability in the industry as well. There was however a degree of nuance to the SSNCo's '*swadeshi*' characteristics. Whilst the ability to hold shares was restricted to non-Europeans as an effort to develop indigenous forms of capitalism, the ships of the SSNCo utilised European officers (Rajendran 1994), presumably because of a lack of adequately trained Indians. This not only shows a degree of pragmatism but also exposes the very real limits which the SSNCo was facing in its attempts to build a 'national' or regional business, particularly in industrial sectors where there was limited local or indigenous expertise, like shipping.

The exact struggle of the SSNCo to compete with the BISNCo is also somewhat unclear, and often only reported through newspapers. A trade war of sorts occurred, and both sides reduced their charges to attract passengers and cargo – something which was obviously easier for the BISNCo to bear as the Tuticorin–Colombo route was only one of its many interests. There were also a number of more underhand practices employed by both sides – Khilnani (2016), based on stories in *The Hindu*, states that the BISNCo started offering free umbrellas and hanging signs with the word '*swadeshi*' on them outside their offices in an effort to confuse Indian customers into buying tickets for their ships. These events meant that the SSNCo became something of a cause célèbre amongst anticolonialists – which is one reason why Aurobindo in Bengal was writing about the Company – and the perception that the BISNCo and the colonial authorities colluded against it was a key source of grievance over the coming months. The *Nadegannadi*, published in Bangalore, then part of the Princely State of Mysore, reported on the 2 May that in the course of the VOC's later trial that a 'secret convention' existed between the BISNCo and railway companies in Ceylon and India to offer discounted train travel for BISNCo passengers. The *Deshabhimani*, an otherwise unremarkable and not notably 'seditionist' newspaper published in Guntur in the Telugu-speaking north of Madras Presidency, wrote in the aftermath of the riots on the 26 March 1908 that:

> in the commercial rivalry that exists between the native and the European Steam Navigation Companies in [Tinnevelly] district, the authorities are always favouring the latter, and have only been waiting for an opportunity to punish the popular leaders, who were influencing the mill employees of the European companies. (MNNR 1907–1908, Cambridge)

Alongside the SSNCo, *swadeshi* activism was becoming prevalent across Tinnevelly district. As the excerpt from the *Deshabhimani* aforementioned hinted at, there was also a wave of industrial unrest in the mills of Tuticorin taking place. Events surrounding the SSNCo culminated in late February and early March 1908 with the workers at the Coral Mills, a British-owned cotton mill in Tuticorin going on strike over pay and conditions, with rumours circulating that a *swadeshi* mill was due to start operation. VOC, along with fellow *swadeshi* speaker Subramania Siva, spoke on the beach at Tuticorin encouraging the establishment of other *swadeshi* enterprises, as well as pointing out various examples of the supposed injustices of foreign rule – from economic hardship through to the non-prosecution of Europeans who had killed Indians. These beach front meetings occurred almost daily throughout February, with around a 1000 people attending on 17 February (Home Political, Branch A June 1908, No. 95, NAI). More mundane acts of resistance were occurring in the town, with *The Hindu* on the 5 March reporting acts of boycott which included barbers refusing to shave individuals who were not pro-*swadeshi* (MNNR 1907–1908, Cambridge). By the 26 February around 3000 people were present at the beachfront meetings, and it was shortly after this that 200 workers at Coral Mills went on strike. Extra police were drafted into the town. Although the town was quiet, there were reports of stone throwing at Europeans and some acts of intimidation to use the SSNCo rather than the BISNCo.

Events carried on escalating, and marches and processions were occurring daily in Tuticorin. On the 7 March officials were reporting that VOC advised the procession:

> to go without sticks. If anything went wrong they must be the injured party and not the offending party. All who wished to have *swaraj* should write the word on their walls or plant a flag on their houses. Every one [*sic*] who is against *swadeshi* should be boycotted. *Swadeshi* boycott and avoiding of the law courts were the weapons to attain *Swaraj*. (Home Political, Branch A June 1908, No. 95)

The tactics of 'passive resistance' were obviously well established here, and, as noted above, we can see that the language of *Swadeshi/Swaraj* was beginning to have cultural effects – in this case, the use of flags and paintings to alter the urban form of Tuticorin and make the anticolonial visible in its streets. By this point, the strike at Coral Mills had been ended, with the workers reportedly being awarded pay-rises of 50 per cent. VOC was involved in the negotiations at Coral Mills on behalf of the strikers, and was able to secure concessions for them. This was important as it showed that upper-class and -caste actors like VOC would actually act on behalf of those seemingly 'below' them in the caste and class hierarchies which were present and to develop a 'national' spirit of solidarity against their minority rulers – something which the *Swadesamitran* of 27 February was vocal in calling for (MNNR 1907–1908, Cambridge). On 13 March, Aurobindo Ghose argued in *Bande Mataram* that the 'Tuticorin Victory' of the strikers was equal to the activities of Gandhi and the Indian community in the Transvaal (Ghose 2002).

This moment of victory was short-lived for VOC and his associates. Having travelled to the district capital of Tinnevelly, VOC, Siva and a number of other *swadeshi*sts began a similar pattern of organising meetings in public spaces, in this case the dry bed of the Thamirabarani river. The *swadeshi*sts organised a meeting and public march to celebrate the release of the extremist leader Bipin Chandra Pal from prison. The march took place on 10 March despite permission for it being refused by the local authorities, and as a result, Pillai, along with his co-organiser, Subramania Siva, was arrested on remand pending trial (Home Political A, Nos. 79–87, April 1908, NAI). Word of the arrests, and their resultant detention without bail, spread and violence broke out on the 13 March with Europeans and their businesses being targeted, and the Municipal Office, Post Office, and Police Station were looted and burnt (Home Political A, Nos. 79–87, April 1908, NAI). As the situation degenerated, the Tinnevelly district Collector L.M. Wynch, ordered the police to open fire, killing four 'rioters' and wounding more. Whilst Tuticorin had been quiet after VOC and Siva left, news of the outbreak of violence in Tinnevelly meant that disorder soon broke out there. It was here that Robert Ashe ordered a crowd to be fired upon, although exactly how is unclear. Ashe's action succeeded in quelling the disturbances in Tuticorin.

Given the recent expansion in political actions in Tinnevelly district, the Government of India (GoI) clearly saw the SSNCo as not simply a business venture but a political front for 'seditious' activities by *swadeshists* (Home Department, Political A, June 1908, No. 95, NAI). The attitude of the authorities in general can be summed up by the report of P.P. Sweeting, the District Superintendent of Police in Tinnevelly, to the Inspector General of Madras on 15 March, where he stated:

> [T]here is great excitement over the arrest of these persons [VOC, Siva and one of their associates] which was only done after the absolute necessity for it was evident. To say that the disturbances here and in Tuticorin on Friday was due to the *arrest* of these sedition-mongers is wrong; it may have formed the *excuse* for the outbreak of the mob who have since the middle of February been taught by speeches made in public by these pestilent sedition-mongers to defy authority and to treat Europeans with contempt, but the real reason is the anti-European, otherwise anti-Government, propaganda that the license allowed to seditious speakers has spread amongst the lower classes, who, in India, as well as anywhere else, can easily be excited to violence. (Home Political, Branch A April 1908, Nos. 79–87, NAI)

The contempt of the coloniser for the colonised is clear here, as well as the need to treat the rebellious or unruly colonised subject with violence – elsewhere in the same report, Sweeting argues that the 'dangerous rabble' had been 'cowed' by the 'severe lesson' they had been taught. It was this punitive attitude which led to the decision to make an example of the organisers. In the subsequent trial, Pillai was sentenced to two life terms in prison, and Subramania Siva to 10 years transportation. The incredibly harsh nature of these sentences provoked further outcry, much of it directed at the District Collector, LM Wynch.

Wynch's actions in opening fire on the unarmed crowd – especially a report that the violence started because he had struck an unarmed shop-owner in the face for displaying a banner in support of Bipin Chandra Pal – as well as his unrepentant attitude after the riots, meant that he became a target for both anticolonialists and the authorities alike. Indeed, the GoI was unconvinced by much of Wynch's testimony in his reports after the riots in Tinnevelly (Home Political, Branch A April 1908, Nos. 79–87, NAI). For the public, the scepticism towards Wynch was combined with the prevailing belief that the authorities were colluding with the BISNCo. The *Patriot*, published in Vellore, went so far as to claim on 31 March that Wynch was a type of 'Deus ex Machina' who arrived to save the situation for the BISNCo, but instead had 'set the prairie on fire' (*The Patriot*, March 31st 1908, MNNR). Elsewhere, the *Liberal* of Madras wrote an open letter to Wynch in March 1908, where it stated:

> I strongly advise you to resign your appointment, or ask for a transfer or long leave. Your order to shoot has not enhanced your reputation as a humane ruler. Tinnevelly has been the grave of the reputation of many civilians; and you are the last mariner who has been wrecked upon the rocky coast of Tinnevelly. To add to these misfortunes, you have apparently sanctioned the prosecution of the most prominent *Swadeshi* preacher and lawyer and five others under the ill-fated security chapter, and afforded additional stimuli to the unrest in your district. Though I am inclined to congratulate you on your having converted Tinnevelly into a district of Eastern Bengal, I tremble to consider what disastrous consequences it will have on the policy of Government. (*The Liberal*, 22 March 1908, MNNR)

Despite the worst of the violence occurring in Tinnevelly, about 36 miles inland of Tuticorin, the maritime nature of the district, and indeed the inspiration of the SSNCo as a maritime venture, is clear to see here. The perception of Bengal as a troublesome province is also important in this relatively moderate, English-language, Indian-produced paper. Catering to the educated classes, the *Liberal's* emphasis on maintaining a stable order, and in engaging in Government-style debates as responsible citizens, not revolutionaries, is important to bear in mind – the letter goes on to discuss how best to intervene in forcing a change of policy using specific legislation as a guideline. However, it shows how important events in Tinnevelly district were if even moderate newspapers were criticising Wynch's actions.

Questions about how fairly Pillai was being treated during the trial dominated the press across India in the immediate aftermath of the trial, and eventually, VOC's sentence was commuted to six years of hard labour, and after protests and petitions eventually ended up serving a term of imprisonment until December 1912 (More 2013). The SSNCo, the riot in Tinnevelly, and Pillai's trial after it made him and *Swadeshi*sm important news across India, and the case became a cause for both *swadeshi*sts and for more moderate observers, and he still remains a recognised 'freedom fighter', especially in Tamil Nadu. The oil press which VOC allegedly worked as hard labour during his imprisonment is now on display in the Gandhi Mandapam in Guindy in South Chennai. He was also the subject of a film, *Kappalottiya Thamizhan* (The Tamil who Launched a Ship), starring the Tamil movie star Sivaji Ganesan in 1961.

Maritime Anticolonialisms

The SSNCo had by the time of the uprisings of March 1908 hit severe financial difficulties. The SS *Gallia* had a technical problem which meant that it had been docked in Colombo since September 1907, and it was alleged that European officers on both the *Gallia* and the *Lawoe* had been defrauding the company over coal supplies (Rajendran 1994, pp. 100–101). VOC's focus on '*swadeshi*' activism rather than book-keeping and his arrest meant that the company foundered and was largely non-existent from the summer of 1908, although it was not formally wound up until 1911.

As we shall see in later chapters and as was indicated in the section on maritime geographies previously, maritime spaces are increasingly recognised as important networks for the dissemination of anticolonial ideas during the early twentieth century. However, the SSNCo provides a different maritime space to the spatially extensive networks which were provided by shipping liners and shipping labour (Hyslop 2009a, 2009b). The SSNCo was, as a small, insurgent enterprise, only able to make a commercially negligible intervention into the international shipping system, dominated locally in Tinnevelly as it was by the BISNCo. Investors into the SSNCo were also not as interested in the goals of *swadeshi* as the likes of VOC. *The Hindu* reported on the 9 March 1908 that the Directors had passed a resolution directing VOC not to organise or become involved in political organising (The Hindu, 9 March 1908, MNNR). The SSNCo was, to some, first and foremost a commercial venture, dedicated towards the capitalist development of India, and the generation of profit to cover their investment. Thus, whilst Aurobindo in the *Bande Mataram* could make exaggerated claims about the potential for *swadeshi* steam companies to dominate Indian Ocean trade, the reality for these companies was an almost impossible struggle against imperial capital, as well as contested internal logics about what the first priority of the company was – economic growth/development or independence.

This also opens up ground to think about the messy nature of political ideology at the time of emergent nationalist and anticolonial frameworks. More (2013) claims that VOC was potentially open to communist and other forms of organising, sharing platforms with speakers who espoused these views and not speaking out. However, he was in charge of running a company which explicitly saw the capitalist development of India as a crucial pathway to modernity and independence. This again shows the uncategorisable and emergent forms of politics which intersected with anticolonialism during this era. There is little point in trying to classify VOC as a strict capitalist or communist, as his ideological position is unknown, and was likely never closely aligned with either. Instead, VOC can be seen as a form of nationalist patriot, and as Heehs (2008) has noted of Aurobindo and his associates, these terms had not acquired many of the negative connotations with which we would now associate them. This importantly tells us something about the dynamic ways in which political identities were formed during the *fin-de-siècle* and is something which tells us about how varied the minor politics of anticolonialism could be.

To return to the maritime, the SSNCo also tells us something about the spatial extent of anticolonial activities. Ports and port cities were seen by the colonial authorities as difficult spaces to manage. Even in provincial ports like Tuticorin, which did not have the international linkages to Europe and elsewhere that Pondicherry or Madras had, the port provided a key space for anticolonial organising. This chimes with work on port cities as spaces of radicalism and distinct urban identities (Mah 2014), but the SSNCo proves useful in thinking across the boundary between land and sea. The nature of *swadeshi* organising here can be thought of as archipelagic, as the ideas and tactics visible in Madras Presidency were deeply connected to those in Bengal and elsewhere in India, so it is possible to think about the various places as distinct 'islands' of *swadeshism*/anticolonialism which were connected to each other. These islands are however 'local' spaces, where Tamilian, Bengali and other regional identities shaped the nature of the various struggles. However, it is in the blurring of land and sea where the metaphor of the archipelago becomes most useful. The protests related to the SSNCo which were the most visible occurred not on board its ships whilst they were at sea – for the most part, travel on board the ships of the SSNCo seems to have been straightforward and trouble-free,[1] and it was through the ideology of its existence that the SSNCo provided a challenge to British rule over South Asia. However, the violent protests which erupted around Tuticorin and Tinnevelly in March 1908 show how the SSNCo's presence stretched into the hinterland of the port city and beyond. The SSNCo acted as an example of what could be possible for Indians as owners and developers of indigenous capital, and the alleged attempts by the BISNCo and other 'British' or 'foreign' interests to disrupt the *swadeshi* enterprise, whether true or not, played into the long-established repertoire of grievances that saw the British as exploiting and draining India's wealth. This also intersected with the ability of VOC and his fellow revolutionaries like Subramania Siva to draw lines of similarity across industries and spur the disputes in the Mills of Tuticorin, which in this case ended relatively successfully for the workers through the negotiated pay increase.

Spatially then the interconnection between land and sea was not a space where the ship became 'invisible' once it had sailed offshore. The regular rhythm of movement of ships between Ceylon and India made the SSNCo part of the cultural landscape of Tuticorin (Anim-Addo 2014) and allowed for the transmission of *swadeshi* and anticolonial ideas to the wider population of Tinnevelly district (and potentially to Colombo, although this is beyond the scope of this book). This

---

[1] There is a report of a collision between an SSNCo steamer and another ship which involved the deaths of three people on the 24 July 1907, but this seems to be the only major incident (see Rajendran 1994, p. 91).

is important as it stretches the politics of nationalism beyond the boundaries of the emergent nation-state, but in different ways to something like diaspora space (Brah 1996) which involves travel to a different nation/culture. The ability of the ships of the SSNCo to act as a 'mutable mobile' of *swadeshi*sm meant that the invisible 'lines' of the shipping routes at sea became, for a very short period, anti-colonial and nationalist spaces, and that crucially, the process of doing this opened up political space for dissent on the mainland.

Conclusions

The violence in Tinnevelly district in 1907 and 1908 was, as previously noted, the largest mass event of violence during the *swadeshi*-inspired agitations in Madras Presidency. *Swadeshi* nationalist versions of anticolonialism were undoubtedly 'political' in the Schmittian sense in their attempts to forge an Indian 'nation' which would be diametrically opposed to foreign, British rule. However, this situation was more complex, as the 'landed' nature of anticolonialism has been challenged in this chapter. This is more than simply saying anticolonial activity happened on ships. Instead, it is to recognise, as David Featherstone (2019) does, that political formations and activities in maritime spaces were distinct and productive of a different set of political relations, which often overlapped with, but crucially exceeded landed concerns.

The SSNCo was, to people like Aurobindo who formed the intellectual wing of the *Swadeshi* Movement, part of an Indian Ocean economy which an emergent India could play a crucial role in. For others, including VOC, the SSNCo was a much needed restorative to Indian cultural and economic values and needs in opposition to the current situation. However, it also shows how organisations like the SSNCo struggled with the limits of industrial expertise available to them – the ships themselves did not involve a renegotiation of any organisational forms, and the shipping company was not promoting some radically 'other' political vision – economic development under conditions of industrialisation was the ultimate aim. The SSNCo then is an example of how *swadeshi* concerns changed when they encountered the sea. In some ways, they continued to deploy the central tropes of swadeshi activism – boycott of foreign goods/companies, and the development of indigenous labour. However, the politics of the SSNCo intersected with other forms of industrial agitation (such as the strike), but hint at a growing mass level to protests in Madras Presidency with the meetings in Tinnevelly's dried up river bed.

Thus, whilst heavily involved in conventional Political activity, we can already see that this was more dynamic and involved the use of a variety of activities that would exceed the more formal categories of the Cambridge School, although in relatively conventional ways. At the same time, in the use of established tropes and social practices (such as in maintaining standard procedures of running and crewing a ship), the SSNCo and VOC were not 'subaltern' in the strictest sense. As a result, we can see the SSNCo as a situated form of anticolonial politics which is heavily based on *ressentiment* – the desire simply to contest the rule of the oppressor. However, it is clearly productive of much more than this, and we must continue to explore where this emergent anticolonial politics in South India would lead.

The following chapters will move further away from this conception of 'the Political' and follow some of the routes that this emergent Tamilian anticolonialism led to. As a result, the next chapter will explore the wider context of the emergence of this politics in Madras Presidency, and how events in Tinnevelly were intimately connected to the lives of the Pondicherry 'Gang'.

References

Amrith, S.S. (2015). *Crossing the Bay of Bengal: The Furies and the Fortunes of Migrants.* Cambridge, MA: Harvard University Press.

Anim-Addo, A. (2014). "The great event of the fortnight": steamship rhythms and colonial communication. *Mobilities* 9 (3): 369–383. https://doi.org/10.1080/17450101.2014.946768.

Balachandran, G. (2006). Circulating through seafaring: Indian seamen, 1890–1945. In: *Society and Circulation: Mobile People and Itenerant Cultures in South Asia 1750–1950* (eds. C. Markovits, J. Pouchepadass and S. Subrahmanyam), 89–130. London: Anthem Press.

Bate, B. (2012). Swadeshi oratory and the development of tamil shorthand. *Economic and Political Weekly* 47 (42): 70–75.

Bate, B. (2013). "To persuade them into speech and action": oratory and the tamil political, Madras, 1905–1919. *Comparative Studies in Society and History* 55 (1): 142–166. https://doi.org/10.1017/S0010417512000618.

Bayly, C.A. (1986). The origins of swadeshi (home industry): cloth and Indian society, 1700–1930. In: *The Social Life of Things: Commodities in Cultural Perspective* (ed. A. Appadurai), 285–321. Cambridge: Cambridge University Press.

Brah, A. (1996). *Cartographies of Diaspora: Contesting Identities.* London: Routledge.

Brennan, L. (1998). Across the Kala Pani: an introduction. *South Asia: Journal of South Asian Studies* 21 (sup001): 1–18. https://doi.org/10.1080/00856409808723348.

Chakrabarty, D. (2010). The home and the world in Sumit Sarkar's history of the Swadeshi movement. In: *The Swadeshi Movement in Bengal, 1903–1908.* Kindle Edi. Ranikhet: Permanent Black.

Chandra, B. (1965). Indian nationalists and the drain, 1880–1905. *Indian Economic and Social History Review* 2 (2): 103–144.

Chandra, B. (2012). *The Writings of Bipan Chandra: The Making of Modern India: From Marx to Gandhi.* Hyderabad: Orient Blackswan.

Chandra, B., Mukherjee, M., Mukherjee, M. et al. (1989). *India's Struggle for Independence.* New Delhi: Penguin.

Chari, S. (2019). Subaltern sea? Indian Ocean errantry against subalternisation. In: *Subaltern Geographies* (eds. T. Jazeel and S. Legg), 191–209. Athens, GA: University of Georgia Press.

Davies, A.D. (2013). Identity and the assemblages of protest: the spatial politics of the Royal Indian Navy Mutiny, 1946. *Geoforum* 48: 24–32. https://doi.org/10.1016/J.GEOFORUM.2013.03.013.

Davies, A.D. (2014). Learning "large ideas" overseas: discipline, (im)mobility and political lives in the Royal Indian Navy Mutiny. *Mobilities* 9 (3): 384–400. https://doi.org/10.1080/17450101.2014.946769.

Davies, A. (2019). Transnational connections and anti-colonial radicalism in the Royal Indian Navy Mutiny, 1946. *Global Networks*.

DeLoughrey, E. (2001). "The litany of islands, the rosary of archipelagos": Caribbean and Pacific Archipelagraphy. *Ariel: A Review of International English Literature* 32 (1): 21–51.

Dittmer, J. and Waterton, E. (2018). "You'll go home with bruises": affect, embodiment and heritage on board HMS Belfast. *Area* https://doi.org/10.1111/area.12513.

Featherstone, D.J. (2009). Counter-insurgency, subalternity and spatial relations: interrogating court-martial narratives of the Nore Mutiny of 1797. *South African Historical Journal* 61 (4): 766–787.

Featherstone, D. (2015). Maritime labour and subaltern geographies of internationalism: black internationalist seafarers' organising in the interwar period. *Political Geography* 49: 7–16. https://doi.org/10.1016/j.polgeo.2015.08.004.

Featherstone, D. (2019). Reading subaltern studies politically: histories from below, spatial relations, and subalternity. In: *Subaltern Geographies* (eds. T. Jazeel and S. Legg), 94–118. Athens, GA: University of Georgia Press.

Ghose, A. (2002). *Bande Mataram: Political Writings and Speeches 1890–1908. Complete Works of Sri Aurobindo*, vol. Vols 6 and 7. Pondicherry: Sri Aurobindo Ashram.

Gilroy, P. (1993). *The Black Atlantic: Modernity and Double Consciousness*. London: Verso.

Goswami, M. (1998). From *Swadeshi* to *Swaraj*: Nation, economy, territory in colonial South Asia, 1870 to 1907. *Comparative Studies in Society and History* 40 (4): 609–636. https://doi.org/10.1017/S0010417598001674.

Goswami, M. (2002). Rethinking the modular nation form: toward a sociohistorical conception of nationalism. *Comparative Studies in Society and History* 44 (4): 770–799.

Goswami, M. (2004). *Producing India: From Colonial Economy to National Space*. Chicago: University of Chicago Press.

Gray, S. (2017). Fuelling mobility: coal and Britain's naval power, c. 1870–1914. *Journal of Historical Geography* 58: 92–103. https://doi.org/10.1016/J.JHG.2017.06.013.

Hasty, W. (2011). Piracy and the production of knowledge in the travels of William Dampier, c.1679–1688. *Journal of Historical Geography* 37 (1): 40–54. https://doi.org/10.1016/j.jhg.2010.08.017.

Heehs, P. (1993). *The Bomb in Bengal: The Rise of Revolutionary Terrorism in India, 1900–1910*. Oxford: Oxford University Press.

Heehs, P. (2008). *The Lives of Sri Aurobindo*. New York: Columbia University Press.

Hofmeyr, I. (2007). The Black Atlantic meets the Indian Ocean: forging new paradigms of transnationalism for the Global South – literary and cultural perspectives. *Social Dynamics* 33 (2): 3–32. https://doi.org/10.1080/02533950708628759.

Hofmeyr, I. (2012). The complicating sea: the Indian Ocean as method. *Comparative Studies of South Asia, Africa and the Middle East* 32 (3): 584–590.

Hyslop, J. (2009a). Guns, drugs and revolutionary propaganda: Indian sailors and smuggling in the 1920s. *South African Historical Journal* 61 (4): 838–846. https://doi.org/10.1080/02582470903500459.

Hyslop, J. (2009b). Steamship empire: Asian, African and British sailors in the merchant marine c.1880–1945. *Journal of Asian and African Studies* 44 (1): 49–67. https://doi.org/10.1177/0021909608098676.

Khilnani, S. (2016). *Incarnations: India in 50 Lives*. London: Penguin.

Lambert, D. (2005). The counter-revolutionary Atlantic: white West Indian petitions and proslavery networks. *Social & Cultural Geography* 6 (3): 405–420. https://doi.org/10.1080/14649360500111345.

Linebaugh, P. and Rediker, M. (2000). *The Many-Headed Hydra: Sailors, Slaves and Commoners and the Hidden History of the Revolutionary Atlantic*. London: Verso.

Mah, A. (2014). *Port Cities and Global Legacies: Urban Identity, Waterfront Work and Radicalism*. Basingstoke: Palgrave Macmillan.

Manjapra, K. (2012). Knowledgeable internationalism and the Swadeshi movement, 1903–1921. *Economic and Political Weekly* 47 (42): 53–62.

More, J.B.P. (2013). *Indian Steamship Ventures, 1836–1910*. Pondicherry: Leon Prouchandy Memorial Sangam.

Naoroji, D. (1901). *Poverty and Un-British Rule in India*. London: Swan Sonnenschein.

Ogborn, M. (2005). Atlantic geographies. *Social & Cultural Geography* 6 (3): 379–385. https://doi.org/10.1080/14649360500111261.

Ogborn, M. (2008). *Global Lives: Britain and the World, 1550–1800*. Cambridge University Press.

Ong, C.-E., Minca, C., and Felder, M. (2014). Disciplined mobility and the emotional subject in Royal Dutch Lloyd's early twentieth century passenger shipping network. *Antipode* 46 (5): 1323–1345. https://doi.org/10.1111/anti.12091.

Osterhammel, J. (2014). *The Transformation of the World: A Global History of the Nineteenth Century*. Princeton, NJ: Princeton University Press.

Pearson, M. (2003). *The Indian Ocean*. London: Routledge.

Peters, K. (2010). Future promises for contemporary social and cultural geographies of the sea. *Geography Compass* 4 (9): 1260–1272. https://doi.org/10.1111/j.1749-8198.2010.00372.x.

Peters, K. (2018). *Sound, Space and Society: Rebel Radio*. Basingstoke: Palgrave Macmillan.

Pugh, J. (2013). Island movements: thinking with the archipelago. *Island Studies Journal* 8 (1): 9–24.

Pugh, J. (2016). The relational turn in island geographies: bringing together island, sea and ship relations and the case of the Landship. *Social & Cultural Geography* 17 (6): 1040–1059. https://doi.org/10.1080/14649365.2016.1147064.

Rajendran, N. (1994). *National Movement in Tamil Nadu, 1905–1914: Agitational Politics and State Coercion*. Madras: Oxford University Press.

Ramaswamy, S. (2010). *The Goddess and the Nation: Mapping Mother India*. Durham, NC: Duke University Press.

Ryan, J.R. (2006). "Our home on the ocean": Lady Brassey and the voyages of the Sunbeam, 1874–1887. *Journal of Historical Geography* 32 (3): 579–604. https://doi.org/10.1016/J.JHG.2005.10.007.

Sanyal, S. (2008). Legitimizing violence: seditious propaganda and revolutionary pamphlets in Bengal, 1908–1918. *The Journal of Asian Studies* 67 (3): 759–787. https://doi.org/10.1017/S0021911808001150.

Sarkar, S. (2010). *The Swadeshi Movement in Bengal 1903–1908*, 2e. Ranikhet: Permanent Black.

Steinberg, P.E. (2013). Of other seas: metaphors and materialities in maritime regions. *Atlantic Studies* 10 (2): 156–169. https://doi.org/10.1080/14788810.2013.785192.

Vannini, P., Baldacchino, G., Guay, L. et al. (2009). Recontinentalizing Canada: arctic ice's liquid modernity and the imagining of a Canadian archipelago. *Island Studies Journal* 4 (2): 121–138.

Venkatachalapathy, A.R. (1994). *Bharathiyin Karathuppadangal "India" 1906–1910 [Cartoons of Bharati: "India" 1906–1910]*. Madras: Narmadha Pathippagam.

Chapter Five
Spacing and Placing Anticolonialism: Pondicherry as a Hub of Radical Nationalist Anticolonial Thought

Introduction

Pondicherry (in French, Pondicherry, and now known as the Tamil Puducherry) is a small city by South Asian standards located on the southeast coast of India, about 150 km south of the city of Madras (now Chennai). Pondicherry was a French territory between 1667 and 1954, so has a longer history of European colonisation than much of 'British' India. Often forgotten or obscured by the imagined British imperial geographies of 'The Raj', French India consisted of the five trading posts or *comptoirs* of Pondicherry, Karaikal, Mahe, Chandernagore, and Yanam, which were separated politically from British India, but were geographically dotted across India. Prior to the Treaties of Paris in 1763 and 1814 which, respectively, first limited and then ended French colonial interests in India, Pondicherry as the capital territory of the five *comptoirs* was a useful space for constituting a French colonial presence in South Asia. Even after 1814, when British interests dominated South Asia, the five *comptoirs* were imagined as *l'Inde Francaise* in an attempt to inflate the importance of these territorial possessions and to create a uniform and cohesive possession out of the scattered and isolated pockets of land (Magedera 2010). Despite these attempts, *l'Inde Francaise* was a decidedly secondary space in the French imperial order. As Chopra (1992) argues, the French possessions in India were accorded much lesser status than those in Indochina or other parts of the French Empire, and Magedera (2003) points out that the French territories were not possessions, but rather *de facto* concessions which were granted to them by the British.

Geographies of Anticolonialism: Political Networks Across and Beyond South India, c. 1900–1930, First Edition. Andrew Davies.

Despite these limits, as distinctive territories that were set apart from British imperial control, the *comptoirs* also provided opportunities for anticolonial radicals who could use them as bases for planning or as sanctuaries. In the early twentieth century, the two French territories of Chandernagore and Pondicherry provided significant opportunities for revolutionary anticolonialists active in the British provinces and territories of India to remove themselves (to some degree) from British surveillance. Whilst this opened up political opportunities, or at least maintained the possibility of conducting political activity, for those who sought sanctuary, it is important to state that both cities were rooted very much in the socio-cultural milieu of the Indian areas in which they were situated – Chandernagore was (and is) Bengali, and Pondicherry distinctly Tamil.

This chapter serves a twofold purpose. First, it continues to develop the discussion of Tamil forms of culture, society and politics which circulated around the various anticolonial activities taking place in Madras Presidency in the early twentieth century. It explores this in particular through a discussion of the life, including a period of 'exile' in Pondicherry, of the Tamil writer Subramania Bharati. Bharati's life is well known in Tamil-speaking regions of South Asia (including Sri Lanka), but he is comparatively less well known outside them, and barely remembered outside India. As a radical who was a part of the extremist wing of the Indian National Congress (INC), as well as a profound believer in pan-Indian freedom, he provides an important insight into how anticolonial ideas were mobilised. However, his life story has an important spatial element – he was forced into exile in Pondicherry from 1908 to 1918, and the various ways in which his opportunities for activism were curtailed in this time tell us much about the wider failures of the anticolonial movement in Madras Presidency. It is also Bharati's exile that drew together other members of what would come to be called the 'Pondicherry Gang' of radicals to the French enclave. Thus, it was thanks to Bharati that the various individuals that form the heart of this book came to be in the same place at roughly the same time. As a result, the second purpose of the chapter is to place Pondicherry as a hub or nexus of anticolonial activity – albeit for a relatively brief time. This is central to the book's objectives to draw out the complex geographies of anticolonialism that interconnected in and through the Pondicherry, establishing an understanding of the city which will contextualise this book's approach. This is important as it shows how anticolonialists were able to alter and reshape the spatial logics of colonialism, in this case, turning Pondicherry from a provincial backwater into an important revolutionary centre, albeit briefly. This sense of anticolonialism's relationship to wider imperial spatialities is something that I return to in this chapter's closing sections. However, the first section of the chapter continues to develop the socio-political context of Madras Presidency and Pondicherry which began in Chapter Three and continued in the last chapter by focussing on the variety and scope of anticolonial and related activities taking place across South India during 1900 to 1910. This will be further developed by an exploration of the life of Bharati. Bharati's biography intersects with some of the other individuals who form the basis of the other chapters, but his life and exile in Pondicherry tell us important facts about the Pondicherry 'Gang'. As a result, a third section will focus on what the anticolonialists in Pondicherry were doing, and how the Government of India (GoI) and the Government of Madras (GoM) were attempting to find out what was happening in French territory. A fourth section covers the most infamous activity of the Pondicherry anticolonialists – the planning of the assassination of Robert Ashe in Tinnevelly district in the South of the Presidency in 1911. Ashe's death was the only political assassination in South India during the entire Indian Freedom Movement, and this episode exposes the need to resort to violence which was espoused by many in the extremist wing of the INC, which many of the men in Pondicherry belonged to. A final section discusses how this chapter's 'placed' account of anticolonial activity helps to expose the situated and relational geographies of anticolonialism.

The Emergent Spaces of Anticolonial Politics in Madras Presidency and Pondicherry

The exact territory of the French *comptoir* of Pondicherry (as opposed to the other four *comptoirs*) was itself made up of a small set of non-contiguous enclaves on the 'Coromandel' coast of South East India. The largest of these enclaves was made up of the city of Pondicherry itself, but there were a series of smaller settlements which were still French territories dotted around the city's immediate surroundings. Despite being the notional 'capital' of *l'Inde Francais*, '[b]y the early nineteenth century, the French Establishments were reduced to small, undefended, politically and geographically inconsequential territories' (Ravi 2010, p. 385). Its backwater status meant that when asked in the 1930s about the city when he arrived in it in April 1910, Aurobindo Ghose described it as 'absolutely dead' (cited in Heehs 2008, p. 218). The surrounding British territory of the Madras Presidency was a different story. As noted in Chapter Three, Madras Presidency has often been characterised as less engaged with the nationalist struggle than other territories. However, during the tensions within the INC between the 'Moderates' and 'Extremists', Madras Presidency was involved in a number of ways. *Swadeshi* activities had spread across all of India following the Partition of Bengal in 1905, and in South/Tamilian India, this occurred in ways which had interesting consequences and are indicative of the growing range of repertoires of contention (McAdam, Tarrow and Tilly 2001) which anticolonialists were using and developing. This is important as it shows how the nationalist politics emerging in India at the time intersected with other political forms, but crucially how these political forms were spatialised.

First, one of the most important ways by which anticolonial nationalist ideas travelled geographically was through various speaking tours, often of the more radical/Extremist members of the INC. For instance, Bipin Chandra Pal (who, alongside Lala Lajpat Rai and Bal Gangadhar Tilak was one of the three extremist INC leaders who were known by the epithet 'Lal, Bal, Pal') toured South India in 1907, going through both Tamil- and Telegu-speaking areas of Madras Presidency. Speeches he gave in Madras were collected and printed in Madras (in English) by the Irish Press in the city (Pal 1907). From this publication, it is clear that notable nationalists in Madras were in attendance, such as G. Subramania Iyer, who edited the firebrand Tamil-language newspaper the *Swadesamitran*. Speaking for hours at a time and being greeted with 'enthusiastic cheering' (Pal 1907, p. 29), Pal and others spoke most often about *swadeshi* and *swaraj*, emphasising the 'extremist' view that full independence from British rule was the only possible solution and moderates like Dadabhai Naoroji were mistaken in calling for more reformist moves to gain more autonomy for India within the British Empire. They also emphasised *swadeshi* tactics, such as explaining and justifying the boy-cott of British and Foreign goods and industries. Speaking tours then provided a clear way of disseminating nationalist rhetoric and ideas directly to audiences. These gatherings were relatively elitist, as the list of attendees would suggest, and gatherings were small scale with tens of people in attendance, however, this was beginning to change, especially as nationalist ideas began to circulate more broadly in South Indian society.

As well as nationalism travelling through speaking tours, an additional set of spaces were the more general public meetings which began to emerge as an important venue for further dissemination of ideas. Often, these took place within any open space that could be found within the various towns and cities of Madras Presidency as we saw in the last chapter, where dry river beds in Tinnevelly allowed for public meetings. In Madras itself, the largest open space was Marina Beach, the huge beachfront that stretches for kilometres along the majority of the city's urban form, and remains today the heartbeat of the city's urban sociality. In 1908, Subramania Siva and Subramania Bharati, two leading Tamil nationalists and newspaper editors announced from a spot on the beach that it was to be called 'Tilak Ghat' after Bal Gangadhar Tilak, another member of the 'Lal Bal Pal' triumvirate (and whose ideas on violence and the political in South Asia were discussed in Chapter Three). These activities ensured that Marina Beach became known as a politicised space in the city throughout the independence movement. In April 1919, the first mass mobilisation against the Rowlatt Act reforms – which were the product of the recommendations of the Rowlatt 'Sedition Committee' Report of 1918 into nationalist and anticolonialist activities and allowed for indefinite detention and incarceration without trial – attracted around 100,000 people on a march around the city with speeches from Presidency and All India scale speakers. Huge loudspeakers carried the messages of the speakers, but they struggled to be heard given the size of the crowd. The status of the beach as a venue to allow public political gatherings meant that Marina Beach continued to attract huge crowds: on 20 March 1919, Mohandas Gandhi spoke on Marina

Beach to announce the Non-cooperation movement, and when he spoke again at Tilak Ghat in 1921, popular memory says that 500,000 people were present.

These huge events of 1919 and after are estimated by Bate (2013) to have been four or five times larger than previous meetings, which prior to the 'Gandhian' phase of the independence struggle were smaller and tended to attract a more middle class, educated or student demographic. It is these earlier meetings and protests which are directly relevant to the subjects of this book. Concerns about these gatherings were present in official circles at both the Presidency and the GoI levels. On 25 June 1907, the GoI wrote to the Chief Secretary of Madras instructing them that police should 'attend openly all political and quasi-political meetings in such force as will frustrate any desire to molest them; and that one or more of the party should openly take notes of the proceedings' (GO Nos. 1407–1408, 10/08/1907, TNA). Despite these measures, by 1908, it was clear that *swadeshi* and other anticolonial activities were becoming increasingly visible in urban spaces like Madras and that these extended beyond the open spaces like Marina Beach and were becoming visible in the streets. In March, H.F. Wilkieson, the Commissioner of Police in Madras, reported to J.N. Atkinson, the Acting Chief Secretary to the GoI, that speeches were being made around Moore Market in Madras almost daily (GO No. 1729, 29/12/1908, TNA). Despite the increasing visibility of protests, police and other officials were extremely uncertain about what effects these *swadeshi* practices would have in Madras (and elsewhere).

Attempts were made to ban or refuse permission for marches and other public speeches, but this air of uncertainty about what would actually happen, as well as an inability to effectively police such events, shines through in the official discussions, with Madras Police asking for clarification or guidance about what the GoI's position was. In March 1908, M.P.T. Acharya applied for permission to use music and fireworks in a procession to commemorate the release of Bipin Chandra Pal from prison. This was refused by Wilkieson, but the march happened anyway on the 9 March. Wilkieson reported to the GoI:

> processions start[ed] from different parts of the city and proceeded towards the South Beach where a public meeting was convened. The processions were orderly till they reached the Victoria Hostel where music commenced and used till they reached the South Beach. It was considered inexpedient to interfere at this stage as the processionist[s] were defiant and in an excited state. The Police did not, therefore, interfere with a view to avoid any disturbance.
>
> After the Procession met on the foreshore of the South Beach two of the speakers named Subrahmanya Bharati [*sic*] and Ethiraj Surendranath Arya in the course of their speeches said that in defiance of the Commissioner's orders they used music and that the audience should take an oath that day that they must be within the legal bounds of the law as far as it did not interfere with their natural rights but when it did so they must infringe the same and break the law. (GO No 1729, 29/12/1908, TNA)

Such flagrant and open defiance of order was obviously concerning to the likes of Wilkieson, but it was also clear that the colonial authorities did not have much understanding of what was happening in Madras and at political meetings more generally. In a note responding to Wilkieson's letter, H.F. Bradley of the Madras Legislative Council attempted to respond with a degree of pragmatism:

> I think it will be much the best thing to take no notice of the breach of the Commissioner's orders. There has been some acrid criticism of the Police in the *Madras Times*, but I haven't heard that any one was damaged or hurt on Monday, except that one lady was discomposed by a restive horse, and someone else had a flag waved in his face. (Side note – A brick was thrown at Mrs. Gunn – H.Bradley, 13/03/08) On the other hand, the crowd went home in a peaceful and orderly manner; and it would be gratuitous fussiness of [us] to launch a prosecution because music was played for a few hundred yards.
>
> The crowd, largely reinforced by the spectators at a football match on the island which had just been played (at the Presidency College – Chiefly students: H. Bradley, 13/03/08), was bigger than the police expected, and they were unable to enforce the prohibition of music, and they exercised a wise discretion in not interfering. If they had interfered and there had been a scuffle the result might have been serious. As it is the papers have said very little about the procession, and except the *Madras Times* I haven't seen that any one has grumbled at the police.

This gives a sense of the dynamic nature of colonial urban spaces and the difficulty of managing them effectively (on this, see also Legg 2007). Whilst keen to downplay the level of disturbance, especially in their correspondence to their superiors in the GoI in Simla/Calcutta, officials in Madras did have fears that *swadeshi* activism would encourage unrest, and concerns about unruly behaviours were keen. Thus, whilst it is noted that nothing major happened, 'rowdy' behaviour amongst the 'native' population had occurred, including the scaring of one English lady's horse. The fact that political marches in urban areas could enrol, even if by accident, larger crowds from lawful gatherings like football matches was also concerning. Criticisms from the *Madras Times*, at that time a British-owned and staffed newspaper with a liberal outlook, would also add to the authorities desire to be cautious for fear of provoking an outcry, but it was clear that tensions were increasing.

Here it is useful to draw upon David Arnold's (2019) chapter on reading the urban space of the street as a subaltern space. Arnold here is attempting to address the relative lack of work on urban areas as spaces of subaltern political activity (for exceptions see Gooptu 2001; Legg 2007). In particular, the aforementioned extracts show how the street in urban areas was a social and political space that, despite the belief in the colonial city as the most orderly and obvious space of imperial pageantry and control, was also a space of life (both human and non-human) and activity. Thus, whilst it is worth remembering that 'the street was more a site of the emerging political hegemony of the Indian middle class than of autonomous urban protest' (Arnold 2019, p. 51), especially when we consider the role of middle-class anticolonial nationalists in organising many of these protests in Madras, the diversity and vitality of urban protest must not be forgotten. Thus, in addition to Arnold, it is worth thinking about the urban anti-colonialisms of the football ground, the beach and other 'urban' public spaces beyond the street which facilitated what Guha (1983) called the 'transmission' of the political amongst India's urban masses.

Thus, a major fear of Madras officialdom, as elsewhere in India, was that anticolonial nationalism would lead to a breakdown in the social order as people questioned British rule and defied existing standards of law and order (see, for example Wagner 2019 here). The language of colonial rule is telling here – as militant anticolonialists were not referred to in these terms, but were rather 'terrorists' or 'anarchists' thanks to their adoption of violent tactics from the various other fin-de-siècle revolutionary movements (Brückenhaus 2017). The language of the terrorist or anarchist 'cell' which is commonplace today was also invisible at the time, with the epithet 'gang', indicating the lawless and suppos-edly unthinking nature of the groups of anticolonialists.

Growing fears, alongside the recognition that current attempts to monitor the 'seditionists' were inadequate, led to the development of new forms of surveil-lance beyond those recommended by the GoI in 1907. As Bate has argued exten-sively, the practices of oration that emerged at this time have had profound effects upon how politics is performed and oralised in rallies across Tamilian India until the present (Bate 2009), but in a more mundane way, it also led to the emergence of vernacular forms of shorthand so that officers could keep up with the speakers (Bate 2012). It is notable that these public meetings shared many similarities with the public speaking practices that emerged from Tamil religious reform move-ments in the mid-nineteenth century (Bate 2005). It is also striking how the newspapers often referred to *swadeshi*-speaking tours or other forms of public speaking as 'preaching' (MNNR 1907–1908). This serves to show not only how the anticolonial domain of the political was, in Madras and elsewhere in the Presidency, articulated through vernacular and cultural distinct forms which not only were different to the rest of India but also how, after Pandian (1995), the political was always more than simply a clientelist jockeying in Madras Presidency.

Public meetings and street protests have always been important spaces of dis-sent, but equally important was the emergence of newspapers and particularly political cartoons within those newspapers. Besides the well-regarded 'English' newspapers like the *Madras Times*, Madras had been a centre of Indian-led news-paper publishing for some time, notably since *The Hindu* began publishing there in 1878. Again, newspapers and their consumption remain important aspects of the Tamil sociality to the present (Cody 2011), but the Tamil-language press (similarly to other linguistic areas of India) formed a crucial space for anticolonial writers to put forward their arguments. Newspapers could be broadly categorised as British-owned/run, Indian-owned and published in English (such as *The Hindu*) or 'vernacular', which were Indian-owned and written in the indigenous scripts and language of the area. There was a huge growth in the numbers of newspapers across Madras Presidency in the latter half of the nineteenth century, with specific papers catering to specific regions or social groups – for example the mixed-race Anglo-Indian community or Malayali-language communities of the western coast of the Presidency. Again, the colonial authorities were swift to try to monitor these developments, with articles from Indian-owned newspapers, whether in English or 'vernacular' languages, being collected in the volumes of 'Native Newspaper Reports' which were maintained across all British territories and sent along to the GoI in Calcutta or New Delhi. From 1905 onwards, attempts were made to proscribe or shut down newspapers which were troublesome across India. In Madras Presidency, there were a number of newspapers which were deemed especially difficult, most notably G. Subramania Iyer's *Swadesamitran* and Subramania Bharati's *India* (Tam. *Intiya*). A detailed discussion of the evolution of publishing in Madras Presidency is not required here (although, see Suntharalingam 1972, 1974 for details here), but it is important to show how innovative these publications were – which will become clear later.

Following the upsurge in newspapers publishing 'seditious' material (which was not confined to the Tamil-language newspapers discussed earlier), there were nine prosecutions in Madras Presidency up until 1909 (GO No. 44, 12/01/1909, Confidential). In 1910, the GoI introduced the India Press Act of 1910 to curtail and control the press within India and also to stop the importing of seditious journals from overseas (such as the *Indian Sociologist* which were being produced by revolutionaries in India House in London and associated with Bhikaji Cama in Paris – see Chapter Seven). Whilst an All India Act, the 1910 Press Act, would have particularly harsh consequences for some of the radical journals which were circulating in Madras Presidency, particularly those being published in Pondicherry. The Act was widely criticised as draconian in its attempts to limit freedom of speech, especially given the small number of newspapers which were actually deemed to be 'seditious' – Venkatachalapathy (2012) states that only eight publications across the Presidency were prosecuted prior to the *Khilafat* and Non-Cooperation Movements in 1919–1921. As a result, the Act was eventually repealed in 1922 (Venkat Raman 1999).

We have then a good idea about what forms anticolonial activity took in Tamil-speaking areas of India. During the years 1907–1910, Madras Presidency was, whilst not as troublesome to the colonial authorities as other provinces like Bengal, a space where anticolonial, specifically *swadeshi*, agitation was increasingly visible. However, as the previous chapter show, political *swadeshi* activity was a potentially explosive issue and was both innovative and deeply rooted within both Tamil and pan-Indian forms of social and cultural organising. In order to explore this intersection between culture and politics, the next section of the chapter explores the life of one of the individuals who was involved in organising and speaking at these events. This is the Tamil nationalist poet and writer

Subramania Bharati. Bharati is a hugely important figure in Tamil culture, credited with reinvigorating the Tamil language as part of the Tamil cultural renaissance of the early twentieth century, but, as we shall see, he was not only a complex figure whose political activity had dramatic consequences for both himself at a personal level but also shaped anticolonial culture in India's most southern provinces.

The Life and Exile of Subramania Bharati

Chinnaswamy Subramania 'Bharati' Iyer was born in the Tamil town of Ettayapuram in the far south of Madras Presidency in 1882, and by the early twentieth century was already renowned as a poet – he was given the honorific 'Bharati', signifying his importance to the land of *Bharat* (an alternative term for India) after a performance in the local *zamindar's*[1] court whilst he was still a child. Bharati's writing has since become so well known that he is classed as a national poet of India – many of his poems and songs became important to the Freedom Struggle in Tamil Nadu, and to Tamil people more generally (Frost 2006). He is credited with being one of the key figures to revitalise the Tamil language in the twentieth century, so much so that his writings were (controversially) nationalised by the Indian State in 1949 after a long legal battle (Venkatachalapathy 2018). However, during his lifetime, Bharati largely lived in poverty and straightened circumstances, and he died in 1921 from dysentery brought on by prolonged poverty, drug use and ill health, hastened by an unfortunate (and possibly apocryphal) incident where he was attacked by a temple elephant in Mylapore in southern Madras.

Bharati was eccentric to say the least, especially considering the social milieu of Tamil Nadu at the time. As a Brahmin by caste (as the Tamilian 'Iyer' caste name indicates), Bharati was born into middle and upper-class circles – his father worked for the local *zamindar* in Ettayapuram, and he was well educated in both Indian and 'Western' styles. However, he actively worked against his caste background and sought to liberalise many of the orthodoxies of Tamil Brahminism. Whilst many of the books about him are distinctly hagiographic and border on the sycophantic, the likes of Mahadevan (1957) do provide some glimpses of Bharati's character. For example, there is an often repeated story where Bharati performed the rite to elevate a Dalit-Bahujan boy who worked for him to Brahminhood as a demonstration of the ridiculousness of caste boundaries. He was also resolutely pan-Indian and often internationalist in his approach, going so far as to mix different styles of dress that were traditionally seen as

---

[1] A *zamindar* was a form of landed aristocrat, many of whom were recognised as princes (and granted their lands as princely states) if they were loyal to the British during the Permanent Settlement.

belonging to certain religious and communal identities, and so his embodied presence in the streets and world of South India was a potential space of minor political rupture or intransigence (cf. Apter 2018 as discussed in Chapter Two). This space of disturbance/rupture is represented in the Tamil biopic *Bharathi* (Rajasekaran 2000) where during the song *Nirpathuve Nadapathuve* ('The lifeless things that stand still' – which is a song Bharati wrote) his behaviour, including walking alongside his wife Chellamma in public and caring for animals, is seen as outraging orthodox Tamil Brahmin sensibilities. This radical nature was further marked by an addiction to opium which often made Bharati's behaviour seem irrational, but which had debilitating consequences on his health throughout his life.

Elsewhere, Bharati's prose writing could include references to Nikolai Tesla or Giuseppe Mazzini alongside discussions of the Hindu principles of the *vedas*. His radicalism extended to women's rights, especially after a meeting with Sister Nivedita, (the Irish-born Margaret Noble who was a disciple of the Indian guru Swami Vivekananda), at the 1906 Calcutta meeting of the INC. Vivekananda is credited with introducing some of the core tenets of Indian spirituality to a Western audience in the Parliament of the World's Religions in Chicago in 1893 and was the founder of the Ramakrishna system of ashrams. Nivedita, as a disciple of Vivekananda, was not only involved heavily in womens' and girls' education programmes for the missions but was also committed to Indian nationalist politics. Nivedita reportedly gently chastised Bharati for not involving his wife in his political activities (Mahadevan 1957). After this, Bharati wrote openly about the need for gender equality in India in his prose (Bharati 1937). This belief in women's emancipation is tempered somewhat by Bharati's tendency to write about women in paternalist terms, often essentialising certain spiritual behaviours and characteristics as feminine. However, this is reflective of wider attitudes in Tamil and Indian society and must be recognised alongside the profound changes that were occurring in Tamil society and which Bharati's writing was playing a key role. What was fundamental, however, was Bharati's belief in liberal and democratic politics, as well as its intersection with his belief in the primacy of spirituality. Certainly between 1905 and 1911, Bharati was also open to the possibility of revolutionary activity in order to achieve a democratic future. These intersections are summarised well by the piece 'Creed of a Democrat' in *Bala Bharata* in December 1907 (Bharati 1907, p. 1165) where he wrote:

I rebel against all forms of fettering, whether of my body, mind or soul.

I rebel against those that are disloyal to the divinely constituted authority of man's reason and man's conscience.

I rebel against the devil-constituted authority of ignorant priest craft and inflated despotism, and purblind society.

I am a spiritual Democrat. My ideal is to establish among mankind a Democracy, not merely of forms but of the spirit.

Bharati, then, was certainly a radical in thought, but was especially so given the constraints of South Indian society in which he spent most of his life. This radicalism and his belief in 'spiritual democracy' meant that he was an ardent anticolonialist, especially whilst writing for a number of newspapers until c. 1910–1911. He was active in *swadeshi* organising throughout South India and attended the Surat Congress meeting of 1907 where the growing fault lines between extremists and moderates split the INC. He was, as can be seen in the extracts from Madras Presidency judicial documents mentioned earlier, well known to the authorities as a public speaker who often appeared at the gatherings on Marina Beach, and was clearly a potential seditious threat. In 1904, Bharati was approached by G. Subramania Iyer, the founder of *The Hindu*, to work on his paper *Swadesamitran*, founded after Iyer left *The Hindu* in 1898. As Venkatachalapathy (2012) has argued, print and publishing culture at the time was undergoing a profound shift away from traditional forms of patronage towards more public ventures. However, this transition was slow and meant that income for writers like Bharati, who even then was recognised as a visionary and important writer of poetry and prose, was tenuous.

Writing for *Swadesamitran* provided some income for a while, until Bharati became disenfranchised by Iyer's unwillingness to support those extremists who were happy to espouse violence. As a result, in 1907, Bharati started publishing both the *India* and the *Bala Bharatha* (Forceful India) with M.P.T. Acharya (see Chapter Seven) in Madras. Bharati's writing in the *India* in particular, with what Venkatachalapathy (2018) describes as a 'racy' style to its political writing, was combined with the first use of political cartoons in India. These cartoons covered a range of issues and were drawn by unknown artists to cover issues and ideas which Bharati had devised (Venkatachalapathy 2006). As a result, the cartoons covered things like the drain of resources out of India under British rule; the *Swadeshi* Steam Navigation Company (SSNCo); intercommunal and pan-Indian harmony; the importance of India as a territorial unit; representations of important anticolonial figures like Sri Aurobindo, and much more (Venkatachalapathy 1994). Ramaswamy (2010) has argued that these cartoons published in *India* (and in his other publications) also provide important representations of India as '*Bharat Mata*' (Mother India), the now common conflation of the territory of India (*Bharat*) as imbued with female/maternal (*Mata*) characteristics.

Given the *India* in a particular's stridently nationalist tone, it was unsurprising that it swiftly attracted the attentions of the Madras authorities and meant that it was swiftly having its articles translated and descriptions of the cartoons recorded in the Native Newspaper Reports of Madras Presidency. During early 1908, there were a number of crackdowns on *Swadeshi* nationalists – in Madras Presidency, the SSNCo-inspired agitations had ended with VOC's arrest and trial, whilst Bal Gangadhar Tilak was deported to Burma on sedition charges having written in support of two revolutionaries associated with Aurobindo who had killed two

British women by throwing a bomb at a carriage which they thought contained a British official (see Chapter Six). This crackdown and increased willingness to prosecute the writers and editors of newspapers for sedition soon meant that *India* became targeted for prosecution.

Bharati's writing was indicative of wider trends of revivalism in Indian thought and culture. Across India, there were a number of intellectuals who were writing about India's pre-existing knowledge and culture as a form of primordialist-style nationalism against imperialism (see Sen 1993 for a discussion of this in relation to Bengal, for example). Most often, this intellectual recovery is thought of as 'Hindu revivalism' as it often drew upon classical Indian philosophical sources. Gandhi's deployment of Indian spirituality in *Hind Swaraj* is one aspect of this here, but virtually every nationalist figure in the fin-de-siècle was utilising similar arguments about the style and shape of India's pre-colonial culture. Thus, whilst there was undoubtedly a religious aspect to this work, to classify such work as purely 'Hindu' is to downplay the various factors which led to this revivalism. Organisations such as the Theosophical Society tied Indian spirituality into cosmopolitan networks internationally through events like the International Races Congress of 1911 (Holton 2002), and as Bipin Chandra Pal put it in his memoirs, organisations like the Theosophical Society:

> told our people that instead of having any reason to be ashamed of their past or of the legacies left to them by it, they have every reason to feel justly proud of it all, because their ancient seers and saints had been the spokesmen of the highest truths and their old books, so woefully misunderstood today, had been the repositories of the highest human illumination and wisdom. (Pal 1932, p. 425)

This societal turn towards 'ancient' philosophies was at the heart of Bharati's thinking, despite his scepticism towards traditional gender and caste roles. Frost (2006) has argued that in order to understand Bharati's poetry, it is necessary to recognise that *bhakti*, or the practice of devotion to a god, is central to his writing. In his nationalist writing, Bharati's *bhakti* is directed towards India. For example, writing in the Bala Bharata in January 1908 Bharati discusses a particular image which he had been sent of India in its feminine representation as 'Bharat Mata', which was becoming a central image of *swadeshi* nationalism. Describing for the reader the various elements of this particular image, Bharati declares:

> We understand that this is the first [of many] sketches representing the Awakened National Consciousness of India and its working in and through "the better mind of India" for that "Unity in variety" that is the secret of national unity ... [leads to] the practical realisation of the oneness of humanity, and its relative expression in social life, which is the ultimate realisation of the hackneyed expression "Equality, Liberty and Fraternity", the tripartite expression of the one Principle which is the ultimate

goal of the human race viz. Universal Brotherhood. (Bharati 1908, pp. 17–18; see also Ramaswamy 2010 for a detailed discussion of this gendered representation of India)

The somewhat complicated links Bharati is making here between the sacred, diverse but feminine independent India and the goals of Revolutionary France show how he was moving beyond a straightforward 'revival' of India's past and instead was working through a progressive vision of India that was not 'Hindu', but was a pluralist and internationalist in outlook. This approach is key to understanding Bharati's worldview and his progressive politics. Many of his cartoons published in *India* show mixed communal groups of Indians – Hindus, Muslims, Sikhs and others – working together for the cause of independence (Venkatachalapathy 1994), and in short stories, he often wrote about Hindu–Muslim interaction and tolerance (see, for example the short story Railway Junction originally written in Tamil for the *Swadesamitran* in 1920, but published in English as Bharati 2016). Elsewhere, in a poem 'The New Russia' written after the revolution of 1917, Bharati wrote:

> The Great mother Kali's merciful glance
> Fell on Russian land, and lo,
> There was revolution!
> The tyrant [i.e. the Tsar] fell screaming.
> The gods smote their arms in jubilation.
> The demons whose tears dried up in fumes of bitterness
> Perished in sheer grief.
> (Trans. M.L. Thangappa, in Venkatachalapathy
> 2018, p. 186)

Bharati's writing then was always internationalist in outlook and sought to communicate international and political affairs to people through both poetry and prose. This revolutionised Tamil literature by making it more explicitly political, but this writing also provides an insight into the ways in which the 'Indian spiritual' was blended with the 'Western Political' by writers like Bharati. The false dichotomy between spiritual and secular will be examined in more detail in Chapter Six, but here the links to Goddess Kali, the destroyer of evil, also an important Bengali god (the name Calcutta derives from Kali) links Bharati's writing not only to the intellectual centre of *swadeshi* and nationalist organising but also means that the Revolution in Russia can be utilised by Indian nationalists. To be sure, Bharati had only limited knowledge of events in Russia, but the fact that he was drawing upon them to mobilise these arguments through his poetry is indicative of the cosmopolitan worldview he possessed.

In August 1908, acting on a tip-off that he was due to be arrested, Bharati escaped with his wife, children and the printing press for the *India* to Pondicherry.

Somewhat ignobly, Bharati's escape meant that M. Srinivasan Iyengar, the officially named editor of *India*, who was essentially nothing more than a patsy who had been named in an attempt to minimise risk to Bharati and his fellow editor M.P.T. Acharya, was arrested and imprisoned in his stead (GO No. 1103, 11/08/1908, [Confidential], TNA; Home Political, Branch A, December 1908, Nos. 6–14, NAI). Bharati who managed to evade arrest arrived in Pondicherry and escaping with the printing press meant that he could continue to publish the *India*, and shortly after began to publish a second paper – *Vijaya* (Victory). Whilst there were 'civil' exiles present in Pondicherry at this time – usually people seeking to escape debts or the police in British India – Bharati's arrival meant that the town became over the next few years a space of political exile. As a well-known public figure who had spoken and written in a number of radical platforms about the need for a renewal of India as well as outright civil disobedience, Bharati's presence in Pondicherry meant that he was followed by a number of other radicals from both Madras Presidency and further afield. From this point on, and for nearly a decade, the Pondicherry 'gang' became a focus of criminal investigation department (CID) activity for Madras.

The Life and Times of the 'Pondicherry Gang'

Shortly after Bharati, his close associate M.P.T. Acharya arrived in Pondicherry (Acharya 1991). A number of Pondicherry residents were sympathetic to the cause of Indian independence and helped to settle the various exiles. Srinivasacharya, another associate of Bharati's, and other well-known *swadeshi* speakers from Madras gradually arrived in the city over the next few years. In December 1910, V.V.S. Aiyer, a Tamil revolutionary, arrived in Pondicherry having avoided arrest in Paris for his activities as part of the wider European revolutionary movement. Most notably, the Bengali revolutionary Aurobindo Ghose arrived in Pondicherry in April 1910 having caught the French-owned steamship *Dupleix* from Chandernagore, where he had been in hiding since escaping there from Calcutta in February that year. Apart from these figureheads, there were a steady stream of less well-known individuals who did much of the work of moving seditious materials around and getting them across the border between French and British territory. To combat this, there were a number of attempts to regulate the postal system in both Pondicherry and Chandernagore (Home Political B July 1908, No. 40, NAI).

To the GoI, the growing number of exiles in Pondicherry was symptomatic of a wider problem of radicals seeking refuge in 'foreign' territory – the *comptoir* of Chandernagore was equally, if not more, of a concern, given its proximity to the centres of violent *swadeshi* activism in Bengal in this regard (Heehs 1993; Sarkar 2010). The French tolerated the presence of the revolutionaries in Pondicherry and made little active effort to deport them, given that they had not committed any illegal acts (this was a somewhat different story in Chandernagore, where a bomb attempt was made on the Mayor of Chandernagore in April 1908 (Sarkar 2010)), and this ambivalent attitude was generally the case prior to the thawing of Franco–British relations in the run-up to World War One. The British tried on a number of occasions to negotiate with the French to buy or exchange some of these territories (Home, Political, Series A, May 1912, Nos. 28–29), including one proposal to swap them with equivalent territories in West Africa. However, the symbolic importance of *l'Inde Francais* to the esteem of France as a colonial power meant that these moves were met with little traction. It was not until May 1918, well after the peak of Pondicherry's role as an active centre of revolutionary planning (but while Chandernagore remained an important space for refuge), that an agreement was reached for the deportation of revolutionaries, and then only from Chandernagore (Home, Political A, May 1918, Nos. 308–310, NAI).

Whilst there were official blockages to the removal of the revolutionaries in French India, it proved remarkably easy for the British to set up surveillance networks within French territory. The Treaty of Paris in 1814 ensured finally that no French military presence beyond policing was allowed in the various *comptoirs* (Ravi 2010), so there were few grounds by which the French could force a stop to a British presence. In September 1911, the Legislative Council and Judiciary of Madras agreed to second C.C. Longden, a Superintendent of the Madras Police, to Pondicherry to officially monitor the 'anarchists' based there (GOs 1010–1014, Tamil Nadu Archives), and a police presence was maintained until 1938 (Arnold 1986). Whilst there was a diplomatic game to be played in maintaining a semi-official police presence in French territory without offending the French governor (GO No. 408 Judicial (Mis. Confidential) 1912 Dated 13/03/1912, GO No. 1014 Judicial (Confidential) Dated 24/06/1912, both TNA), by July 1912, there were 7 sub-inspectors and 42 policemen being supervised by Longden (GO 1335, TNA). There were also considerable numbers of undercover police, and the sudden appearance of both a revolutionary group and a huge number of police and their informers 'enlivened the life' of previously dull and monotonous Pondicherry according to Madras newspaper *The Standard's* correspondent in 1912 (GO 1335, TNA).

Despite these attempts at control, Pondicherry proved to be, at least for a short time, a viable space for anticolonial activity. For one thing, the port of the city allowed the import of seditious materials ranging from pamphlets and books to firearms. As Hyslop (2009) has noted, ports were insecure spaces of colonial empires, and often allowed the relatively free flow of materials through them as part of the global networks of imperial steamship trade. The Government of India recognised that Pondicherry as a port with direct sailings to Europe made it easier for smuggling to take place, compared to the more remote Chandernagore. Madras Presidency admitted to the fact that arms smuggling was taking place through Pondicherry into the adjacent British district of South Arcot – and whilst it was occurring in small numbers it was almost impossible to stop

(Home Political, Branch A, December 1913, Nos. 15–16, NAI). In 1911, it was reported by the Criminal Intelligence Office that after a raid on the revolutionaries, receipts were found over the course of a year for: 456 copies of *Bande Mataram* (published in Geneva and not to be confused with Aurobindo's publication of the same name in Bengal); 156 of the *Talvar* (Berlin); 124 of the *Gaelic American* (USA); 109 of the *Indian Sociologist* (London and Paris); together with 16 copies of Savarkar's *Indian War of Independence* as well as other radical journals and pamphlets had been imported in to Pondicherry (Home, Political (Deposit), March 1911, No. 12, NAI). Arms were also trafficked into Pondicherry and Chandernagore, and the differing regulation of the postal system in French and British territories meant that there were a number of loopholes which could be exploited – for instance, importing guns by sea to Pondicherry, then sending by land in the French postal system to Chandernagore was identified as a potential way for this smuggling to occur (Home Political (Deposit) April 1910, No. 20, NAI). Pondicherry had become an important hub in a global network of anticolonial revolutionaries, and an important place for these movements and their ideas to gain access to India.

On the other hand, it was harder for revolutionaries to make a living whilst in Pondicherry, and this was largely due to the various attempts to control them financially that were imposed from across the border in Madras (Home, Political Branch B, February 1910, Nos. 143–145; Home Department, Political, Branch B, May 1910, Nos. 191–193, both NAI). So whilst the likes of Bharati could write newspapers in the short term, it was harder for them to receive subscriptions addressed to them, and by 1912, any income which came from publishing newspapers and pamphlets had dried up. This, given the transitions towards a public rather than patronage-based publishing culture noted by Venkatachalapathy (2012) earlier, meant that the revolutionaries in Pondicherry were severely financially restricted. Cyril Longden in a report to the Madras CID in August 1912 summarised some of these difficulties:

> Without being absolutely on their last legs they are distinctly in straitened circumstances. VVS [Aiyer] and [Aurobindo Ghose] are both changing their houses from motives of economy. Bharathi [*sic*] has the greatest difficulty in getting his poems printed at all and a Pondicherry press does not charge prohibitive rates. T.S. Srinivasamurthi has practically nothing and if [Aurobindo] does lend the others money (he has lent [Srinivasacharya] money) he takes extraordinary good care that he gets it back sharp, even if his young men have to wait all day for it. Local support is forthcoming (I know of one Chetty[2] who gives Rs 3 a month) but it is not lavishly paid, and the older folk [in Pondicherry] have set their faces very strongly against their sons associating with the extremists. VVS [Aiyer] has been importing sham jewels which D.S. Madava Rao sends him, and is trying to export skins and poppadoms

---

[2] Chetti/Chettiars is the grouping used by the majority of business castes in South India.

and so make a bit of money by trade. [Aurobindo] of course has money which he gets through the Banks, and his Bengalis spend their time in a reading room and are apparently shining lights at the local games clubs, football and hockey especially, as far as I hear, being their favourites. (GO No. 1335 Judicial Department (Confidential) Dated 21/08/1912, Tamil Nadu Archives)

Despite the increasing ability of the colonial state to monitor the activities of the 'gang' in Pondicherry, it proved remarkably difficult to monitor the exact comings and goings of goods and people, not least because of the limited sources of reliable information. The diary of the British Consul in Pondicherry on 1 September 1910 noted:

When the SS *Dupleix* [the main ship steaming between the French *comptoirs*] went up to Calcutta on the 6th of August a case of revolvers was landed consigned to a native said to contain 40 [guns]. It was not opened by the French Customs at the Port office and was taken into the town. Mons. L Coruet, Engineer, vouches for the truth of this but he was unable to make out the name of the consignee, however the bill of landing of which he caught a glimpse showed 40 revolvers. He dare not report this to the police for fear of being set upon by the natives. My man's endeavouring to ascertain the name. The French customs officials are most despicable, for a little 'douceur' they will allow anything through. (Diary of His Britannic Majesty's Consul at Pondicherry, NAI)

As well as guns, printed material also circulated relatively freely. As Venkatachalapathy (2012) discusses, even after the Press Act of 1910, the Madras Registrar of Books was only a part-time position, with an attached staff of one head clerk, three clerks to read any books, one typist and two peons.[3] The colonial state machinery was therefore incredibly limited and far from panoptic in nature. In the same diary entry quoted earlier, the British Consul reported that copies of the *India*, as well as bundles of a leaflet called *Dharman* were smuggled over the border by some 'Brahmins' without any apparent trouble. As a result, it is hard to know exactly how much material was moved through Pondicherry, simply because there are no accurate records. What is certain is that movements of people and material objects continued – often with revolutionaries adopting disguises to appear as though they were from different parts of India (Home Political A, 1911, Nos. 114–117, NAI).

Given the long period of exile, stretching into years, and for some decades, Pondicherry also became a space of domesticity for the revolutionaries. It was noted earlier how some of the men engaged in playing games in the local sporting clubs. Pondicherry's beachfront proved to be a space where revolutionaries could meet and discuss their plans, and the British Consul reported on 15 October 1910 that Aurobindo came out of 'seclusion' onto the streets of Pondicherry for the first time since he arrived, walking with Bharati along the waterfront (Diary

---

[3] 'Peons' is a term for low-ranking or menial workers employed by many businesses across South Asia.

of His Britannic Majesty's Consul at Pondicherry, NAI; see also Home Political A, 1911, Nos. 114–117, NAI).

Small groups of the 'secret society' would meet in each other's houses. Whilst there was a core of revolutionaries like Bharati who were required to stay in Pondicherry under threat of arrest in British territory, others came and went and did the work of distributing anticolonial pamphlets and literature, as well as smuggling guns. These people often stayed in the houses of the various revolutionaries as necessary, and a degree of fraternal bonding was present. It must also be noted that, for men like Bharati, they also lived with their families in exile. Sadly, there is little known about how the wives and children of the various anticolonialists coped with these circumstances. Bharati's wife Chellamma, according to her granddaughter, provided the inspiration for much of his poetry (Bharati 2003), and she was instrumental in many of the fights over his intellectual property after his death (Venkatachalapathy 2018), but apart from passing references to the hardship in which the family lived (not helped by Bharati's often irrational behaviour, alongside his addiction to opium) there remains little knowledge of the everyday and domestic spaces of the Pondicherry Gang. However, the planning and action that the men and women of the Gang were involved in, often in their houses and homes, led to a significant and violent event.

Revolutionary Violence and the Assassination of Robert Ashe

On 17 June 1911, whilst sitting next to his wife on a train in Maniyachi Junction in Tinnevelly district in the far south of Madras Presidency, the District Collector, Robert Ashe, was shot dead by Vanchi Aiyar, a relatively unknown revolutionary who was from Travancore – a princely state bordering Madras Presidency to the South West. Ashe had been targeted because of his association with the 'Tinnevelly Riots' in 1907, but for which his superior at the time, Collector Wynch, was largely to blame. Having killed Ashe, Vanchi hid in the toilets of the station and was discovered dead shortly after having shot himself (Home, Political, Branch A, June 1912, Nos. 41–68, NAI). On his body, a letter threatening a wider plot to assassinate King George V was found. A massive manhunt followed and amongst his associates a number of political publications were found, including books written by Bharati, as well as pamphlets which could be traced to Pondicherry.

Ashe's death was the only political assassination in Madras Presidency during the entire freedom struggle and elicited widespread condemnation. The Ashe papers, now held at Cambridge University, contain 238 telegrams or letters of condolence sent to Ashe's wife after his death (Ashe Papers, Cambridge). By 29 June, the GoI was writing to the Under Secretary of State for India in the UK stating that Vanchi had clear links to the anticolonialist revolutionaries in Pondicherry. The Governor of the French Territories in India, M. Martineau, wrote to the GoI, recognising that the activities of anticolonialists in French India had played a part in the assassination and that 'I am now obliged to acknowledge that there is amongst European nations an intimate responsibility in this country and that, as you say, there would be an advantage in cooperating to take steps to crush dangerous organisations of which, I still hope, these are but temporary manifestations' (Home, Political, Part B, July 1911, Nos. 41–42, NAI). Thus, some aspects of the *détente* between British and French colonial powers and the increased openness to cross-border surveillance between British and French India can be traced to this assassination.

Ashe's assassination sparked a severe clampdown on the Pondicherry Gang – Superintendent Longden's posting to the city to organise the surveillance of the group started directly after Ashe's death. A network of (often unreliable) informers within the city swiftly provided information to the British, many of whom clearly implicated Bharati, V.V.S. Aiyer and other revolutionaries in the planning of the assassination. The file Home Political A, 1911, Nos. 114–117 contained in the National Archives of India contains a range of testimonies from Pondicherry residents collected by the District Magistrate of Tinnevelly after the murder. From these, it was clearly widespread knowledge that the members of the Gang, led by Iyer, were involved in revolver practice in Pondicherry, which took place every Sunday with around 30 people in attendance. Informers estimate that the 'Gang' had dozens of revolvers in their possession which had been smuggled from France, and most claim that Vanchi and one of his accomplices Nilankanta was trained at these meetings. The revolver practices were often conducted in the gardens of sympathisers in Pondicherry, and occasionally used images of foreigners as targets. The practices were often followed with lectures by the likes of Aiyer and Bharati who emphasised the belief that action, potentially violent, would be necessary to overthrow the British. In further evidence of the fraternal and domestic nature of the Gang's lives in Pondicherry, apparently Vanchi lived with Bharati when he first arrived in Pondicherry at some point in 1910. Nilankanta apparently used to eat at Bharati's house, something which was rare amongst the orthodox Tamil Brahmin community. The evidence between the informers is consistent enough to believe that much of this information is accurate, and on 4 September, it was agreed that eight people in Pondicherry, including Aiyer, Bharati and Srinivasacharya were to have their post intercepted and delivered instead to Longden. Whilst the détente between British and French was not significant enough to lead to deportations, and thus none of the 'Gang' were prosecuted as part of the Ashe conspiracy, the Rowlatt Report of 1918 (Rowlatt Committee 1918) continued to link them to the assassination.

Political violence was a key aim for some members of the Pondicherry Gang, and for some, the stockpiling of weapons was a part of the planning for a wider revolution in India. Given the groups' longstanding connections with the extremist wing of the INC, it is unsurprising that the methods promoted by the likes of Tilak would come to be important to the group. This does, however, further challenge Washbrook's now-disputed claim that 'between 1895 and 1916, scarcely a single anti-British dog barked on the streets [of Madras Presidency]' (Washbrook 1976,

p. 233). Longden reported in a note to Madras that, even in the financially parlous condition, the revolutionaries found themselves in after Ashe's assassination, that Aiyer in particular was reading Von Clausewitz's 'On War' in preparation for the coming insurrection (GO 1335, 1912, TNA). However, the small number of anti-colonialists meant that this was never likely to happen – the *swadeshi* movement in Madras Presidency was confined largely to the educated classes who could read pamphlets and newspapers, and never acquired mass movement characteristics. Ashe's assassination was the final notable action of the *swadeshi*-inspired anticolo-nialists in Southern India – following it, police in Madras Presidency swiftly iden-tified and prosecuted the immediate accomplices of Vanchi. Other radicals were already in jail, and the clampdown on periodicals coming over the border from Pondicherry meant that the ability of the revolutionaries to organise any activity was in terminal decline.

The swift suppression of the core anticolonialists and the minor scale of revo-lutionary political violence in Madras does not mean it is insignificant. Instead, it highlights how closely connected Madras Presidency was to the actions taking place in India and in Europe. The ability of the group to move illicit goods through Pondicherry meant it was a nodal point in this early period of swadeshi activism, and the practices of revolutionary violence which were taking place in Bengal and in Europe (Heehs 1992; Brückenhaus 2017) were known to the group.

Conclusion: Exile and the Place of Anticolonialism

From Bharati's arrival in 1908, the crackdown on the Pondicherry 'Gang' after Ashe's assassination meant that Pondicherry was an active 'hub' of revolutionary activity for less than a five years. The increased surveillance of *swadeshi* activism after Ashe's murder largely disrupted any attempts at organising in Madras Presidency, together with the end of the swadeshi movement elsewhere in India and Gandhi's arrival in India in 1915 meant that Madras Presidency and Pondicherry alike became areas of little concern in terms of anticolonial and nationalist organising. Bharati himself remained in Pondicherry until 1918, when he returned to British India after it became clear that he was not likely to be jailed following the decline of violent anticolonialism in the Presidency. Bharati himself had also mellowed somewhat and had become more closely aligned with Gandhian thought. He spent the final few years of his life trying to eke out a living through publication, including some of the huge volume of writing he had been doing whilst living in Pondicherry. Much of this was not successful in his lifetime (see Venkatachalapathy 2018), and as noted earlier, he died in poverty on 11 September 1921 in Madras.

This chapter's focus on individuals like Bharati and their time in Pondicherry is important to the book's aim as a whole as it shows how anticolonial politics becomes grounded and mobilised through particular places and spaces. Similar narratives could, for example, be constructed around Algiers, Havana, Dar es Salaam or any other city where anticolonial politics came to dominate for a period of time. Pondicherry, however, is particularly interesting given its particular spatial context as a minor enclave of a French colonial power in a region dominated by the UK. The French enclaves in India provided for a short time a strategic sanctuary (Edwards 2010) for anticolonialists to carry out clandestine activities. However, the weakness of the French position in limiting the British presence in its territories, as well as the recognition by the French authorities that the anticolonialists in Pondicherry were equally likely to mobilise against French colonisation, meant that these spaces were only ever likely to be a temporary solution in the back and forth strategies of colonial power and anticolonial resistance.

As I have argued elsewhere (Davies 2017), the exile of Subramania Bharati in Pondicherry is an important example of the complex spatialities of colonialism, and the anticolonial politics of friendship which were created there. The ability to exist in a state of 'exile' within a Tamil cultural space where everyday cultural markers would be well known to Bharati and others like Aiyer and Srinivasacharya, opens up a window into the varied territorial and political spaces of colonialism. In contrast, to Bengali exiles, Tamil social mores were often distinctly unknown – as Bipin Chandra Pal noted in the 1930s, as recently as the 1880s Madras was 'more or less an unknown land to us [Bengalis]' (Pal 1932, p. 386), and as Heehs (2008) describes, when one of Aurobindo's Bengali associates arrived in Pondicherry in 1910 to prepare for Aurobindo's eventual arrival, their inability to speak Tamil, French or English meant that it was incredibly hard to find the offices of the *India*, where it was known that they could ask for assistance. The geography of 'exile' was at once not only a space of hardship and containment for the likes of Bharati but also was a highly networked space which brought 'Indians' together, and also where connections to Europe were arguably much less regulated than they would have been in British India.

The spatially varied freedoms afforded to the citizens and subjects of empire as they moved around the world are well known (Banerjee 2010; Tickell 2011), but spaces like Pondicherry show how anticolonialists worked around and attempted to undermine the restrictions which were placed upon them, and were potentially spatially mobile if necessary. This also exposes the constant need for the British colonial authorities to respond to anticolonialists like the Pondicherry Gang in order to maintain a sense of control for continued British rule in India. This forms part of the constant churn of colonial and anticolonial practises of domination and resistance, as both sides responded to each other, but also is indicative of some of the wider spatial challenges which anticolonial activities forced upon colonial rulers. As Stephen Legg (2009) has argued, in relation to the spatiality of the League of Nations and the GoI, scalar politics are best imagined as a series of effects, and that the interactions between scales and networks of power during the European colonial era were not fixed, despite the desire of the state to claim that they were in order to maintain the system of imperial and colonial rule.

The spatial network of revolutionaries in Europe and elsewhere in India who connected to Pondicherry destabilised, for a short time, the idea of Pondicherry as a backwater of little or no significance to British dominance in India. The desire by the GoI to swap or incorporate various French territories of South Asia shows how important it was to maintain the idea of British domination of South Asia, even if spaces like Pondicherry and Chandernagore were incredibly marginal compared to the rest of India. However, the desire of the French to keep these possessions also shows how symbolically important these marginal spaces were. Thus, the placed and networked politics of the 'Pondicherry Gang' exposes something of the fallibility of colonial power during this era, especially in their struggles to coordinate national- and colonial-scale state apparatuses to international networks of subversion – a theme which we will return to in Chapter Seven (see also Wilson 2016 on the limits of British rule).

However, it is also important to note here the role that these anticolonialists played in reshaping Pondicherry and its image. As noted earlier, Pondicherry was a backwater when the likes of Bharati arrived. Following Aurobindo's arrival – and especially after his turn towards spirituality which will be the focus of Chapter Six – Pondicherry became a centre for distinctly cosmopolitan forms of social and political activity. Today, the city's postcolonial reinvention and place-marketing not only focus on the heritage of the city as a distinctly Francophone colonial space (Jørgensen 2017) but also a distinctly New Age one connected to the legacies of the Sri Aurobindo Ashram and Auroville, the experimental city established by his followers in the late 1960s (Namakkal 2012; Jazeel 2015). Whilst Aurobindo, as we shall come to in the next chapter, is probably the most notable of the Pondicherry 'Gang' today, Bharati's influence on both the city and on wider Tamil culture is also clear. From the statue of Bharati in Bharati park (which is not named after him) in the centre of Pondicherry's Heritage Quarter, the official memorial museum to him in one of the houses he lived in on Eswaran Dharmaraja Kovil Street, or the schools named after him, Bharati's presence within the city still remains visible. Understanding Pondicherry in the present, then, requires a sense of what happened in the city a century ago amongst the subversive gatherings of these revolutionaries.

The story of the 'Pondicherry Gang' then is one of the relational geographies of anticolonialism intersecting in the production of a specific place – to paraphrase Doreen Massey, a place-based, global sense of anticolonialism. Understanding someone like Bharati means that we must first see anticolonialism as culturally contingent – Tamil anticolonialisms were rooted in distinctly Tamil forms of cultural practice, which in Bharati's case involved reinvigorating the Tamil language as a whole so that it could cope with his dynamic political writing and cartoons in the *India*. Thus, whilst sharing many characteristics of the wider swadeshi organising, the specific geographies of Tamil India made it distinct. This chapter then tells us something of a traditional 'geography' of anticolonialism – with political contestation being rooted in a particular place, but routed through that place's connections to elsewhere. However, this only forms one way in which the anticolonial and the political came to be articulated in Pondicherry.

References

Acharya, M. (1991). *Reminiscences of an Indian Revolutionary* (ed. B. Yadav). New Delhi: Anmol.

Apter, E. (2018). *Unexceptional Politics: Obstruction, Impasse and the Impolitic.* London: Verso.

Arnold, D. (1986). *Police Power and Colonial Rule: Madras 1859–1947.* Bombay: Oxford University Press.

Arnold, D. (2019). Subaltern streets: India 1870–1947. In: *Subaltern Geographies* (eds. T. Jazeel and S. Legg), 36–57. Athens, GA: University of Georgia Press.

Banerjee, S. (2010). *Becoming Imperial Citizens: Indians in the Late-Victorian Empire.* Durham, NC: Duke University Press.

Bate, B. (2005). Arumuga Navalar, Saivite sermons and the delimitation of religion, c. 1850. *The Indian Economic and Social History Review* 42 (2): 469–484.

Bate, B. (2009). *Tamil Oratory and the Dravidian Aesthetic: Democratic Practice in South India.* New York: Columbia University Press.

Bate, B. (2012). Swadeshi oratory and the development of Tamil shorthand. *Economic and Political Weekly* 47 (42): 70–75.

Bate, B. (2013). "To persuade them into speech and action": oratory and the Tamil political, Madras, 1905–1919. *Comparative Studies in Society and History* 55 (1): 142–166. https://doi.org/10.1017/S0010417512000618.

Bharati, S. (1907). Creed of a democrat. In: *Kālavaricaip paṭuttaṭṭa pārati paṭaippukaḷ: Iraṇṭām tokuti, 1907 [Chronological works of Bharati: Volume 2, 1907]*, 2007e (ed. S. Viswanathan), 1165–1166. Chennai: Viswanathan, Sili.

Bharati, S. (1908). Bharati Mata, or, the Awakened National Consciousness of India. In: *Kālavaricaip paṭuttaṭṭa pārati paṭaippukaḷ: Iraṇṭām tokuti, 1908 [Chronological works of Bharati: Volume 3, 1908]*, 2002e (ed. S. Viswanathan), 16–18. Chennai: Viswanathan, Sili.

Bharati, S. (1937). *Essays and Other Prose Fragments.* Madras: Bharati Prachur Alayam.

Bharati, S. V. (2003) 'Remembering Chellammal Bharati', *The Hindu.* September 21st.

Bharati, S. (2016). Railway junction. In: *The Tamil Story* (ed. D. Kumar). Chennai: Kindle Edi.

Brückenhaus, D. (2017). *Policing Transnational Protest: Liberal Imperialism and the Surveillance of Anticolonialists in Europe, 1905–1945.* Oxford: Oxford University Press.

Chopra, P. (1992). Pondicherry: a French enclave in India. In: *Forms of Dominance: On the Architecture and Urbanism of the Colonial Enterprise* (ed. N. AlSay), 107–137. Aldershot: Ashgate.

Cody, F. (2011). Echoes of the teashop in a Tamil newspaper. *Language & Communication* 31 (3): 243–254. https://doi.org/10.1016/j.langcom.2011.02.005.

Davies, A. (2017). Exile in the homeland? Anti-colonialism, subaltern geographies and the politics of friendship in early twentieth century Pondicherry, India. *Environment and Planning D: Society and Space* 35 (3): 457–474. https://doi.org/10.1177/0263775816662467.

Edwards, P. (2010). A strategic sanctuary: reading *l'Inde française* through the colonial archive. *Interventions: International Journal of Postcolonial Studies* 12 (3): 356–367.

Frost, C.M. (2006). Bhakti and nationalism in the poetry of Subramania Bharati. *International Journal of Hindu Studies* 10 (1): 150–166.

Gooptu, N. (2001). *The Politics of the Urban Poor in Early Twentieth Century India*. Cambridge: Cambridge University Press.

Guha, R. (1983). *Elementary Aspects of Peasant Insurgency in Colonial India*, 1999e. Durham, NC: Duke University Press.

Heehs, P. (1992). The Maniktala secret society: an early Bengali terrorist group. *The Indian Economic and Social History Review* 29 (3): 349–370.

Heehs, P. (1993). *The Bomb in Bengal: The Rise of Revolutionary Terroriam in India, 1900–1910*. Oxford: Oxford University Press.

Heehs, P. (2008). *The Lives of Sri Aurobindo*. New York: Columbia University Press.

Holton, R.J. (2002). Cosmopolitanism or cosmopolitanisms? The Universal Races Congress of 1911. *Global Networks* 2 (2): 153–170. https://doi.org/10.1111/1471-0374.00033.

Hyslop, J. (2009). Guns, drugs and revolutionary propaganda: Indian sailors and smuggling in the 1920s. *South African Historical Journal* 61 (4): 838–846. https://doi.org/10.1080/02582470903500459.

Jazeel, T. (2015) Matrimandir, Auroville, *Society and Space*. Available at: http://societyandspace.org/2015/08/19/matrimandir-auroville-tariq-jazeel.

Jørgensen, H. (2017). Between marginality and universality: present tensions and paradoxes in French colonial cultural heritage, civilizing mission, and citizenship in Puducherry, India. *Heritage & Society* 10 (1): 45–67. https://doi.org/10.1080/2159032X.2018.1457299.

Legg, S. (2007). *Spaces of Colonialism: Delhi's Urban Governmentalities*. Oxford: Wiley-Blackwell.

Legg, S. (2009). Of scales, networks and assemblages: the League of Nations apparatus and the scalar sovereignty of the Government of India. *Transactions of the Institute of British Geographers* 34 (2): 234–253. https://doi.org/10.1111/j.1475-5661.2009.00338.x.

Magedera, I.H. (2003). France–India–Britain, (post)colonial triangles: Mauritius/India and Canada/India, (post)colonial tangents. *International Journal of Francophone Studies*: 64–73.

Magedera, I.H. (2010). Arrested development. *Interventions* 12 (3): 331–343. https://doi.org/10.1080/1369801X.2010.516092.

Mahadevan, P. (1957). *Subramania Bharati: A Patriot and a Poet: A Memoir*. Madras: Atri Publishers.

McAdam, D., Tarrow, S., and Tilly, C. (2001). *Dynamics of Contention*. Cambridge: Cambridge University Press.

Namakkal, J. (2012). European dreams, Tamil land: Auroville and the paradox of a postcolonial utopia. *Journal for the Study of Radicalism* 6 (1): 59–88.

Pal, B.C. (1907). *Speeches of Sri B. C. Pal Delivered at Madras*. Madras: Ganesh & Co.

Pal, B.C. (1932). *Memories of My Life and Times*. Calcutta: Modern Book Agency.

Pandian, M.S.S. (1995). Beyond colonial Crumbs: Cambridge School, identity politics and Dravidian movement(s). *Economic and Political Weekly* 30 (7/8): 385–391.

Rajasekaran, G. (2000). *Bharathi*. Chennai, India: Media Dreams Pvt. Ltd.

Ramaswamy, S. (2010). *The Goddess and the Nation: Mapping Mother India*. Durham, NC: Duke University Press.

Ravi, S. (2010). Border zones in colonial spaces. *Interventions* 12 (3): 383–395. https://doi. org/10.1080/1369801X.2010.516096.

Rowlatt Committee (1918) 'Sedition Committee Report'. Edited by G. of I. Home Department. Calcutta: Government of India.

Sarkar, S. (2010). *The Swadeshi Movement in Bengal 1903–1908*, 2e. Ranikhet: Permanent Black.

Sen, A.P. (1993). *Hindu Revivalism in Bengal, 1872–1905: Some Essays in Interpretation*. Oxford: Oxford University Press.

Suntharalingam, R. (1972). The "Hindu" and the genesis of nationalist politics in South India, 1878–1885. *South Asia: Journal of South Asian Studies* 2 (1): 64–80. https://doi. org/10.1080/00856407208730666.

Suntharalingam, R. (1974). *Politics and Nationalist Awakening in South India, 1852–1891*. Tucson: University of Arizona Press.

Tickell, A. (2011). Scholarship terrorists: the India house hostel and the "student problem" in Edwardian London. In: *South Asian Resistances in Britain, 1858–1947* (eds. S. Mukherjee and R. Ahmed), 3–18. London: Continuum.

Venkat Raman, V. (1999). The Indian Press Act of 1910: the press and public opinion at crossroads in the Madras Presidency, 1910–1922. *Proceedings of the Indian History Congress* 60: 863–871.

Venkatachalapathy, A.R. (1994). *Bharathiyin Karathuppadangal "India" 1906–1910 [Cartoons of Bharati: "India" 1906–1910]*. Madras: Narmada Pathippagam.

Venkatachalapathy, A.R. (2006). *In Those Days There Was No Coffee: Writings in Cultural History*. New Delhi: Yoda Press.

Venkatachalapathy, A.R. (2012). *The Province of the Book: Scholars, Scribes, and Scribblers in Colonial Tamilnadu*. Ranikhet: Permanent Black.

Venkatachalapathy, A.R. (2018). *Who Owns that Song? The Battle for Subramania Bharati's Copyright*. New Delhi: Juggernaut.

Wagner, K. (2019). *Amritsar 1919: An Empire of Fear and the Making of a Massacre*. London: Yale University Press.

Washbrook, D. (1976). *The Emergence of Provincial Politics: Madras Presidency 1870–1920*. New Delhi: Vikhas Publishing House.

Wilson, J. (2016). *India Conquered: Britains Raj and the Chaos of Empire*. London: Simon and Schuster.

Chapter Six
Envisioning a Spiritual and Cosmopolitan Decolonial Future? Sri Aurobindo's 'Non-political' Anticolonialism

Introduction

Aurobindo Ghose has been something of a marginal presence in the two previous chapters, despite the fact that he is today probably the most well known of the four men around whom the core chapters of this book orbit. Aurobindo's writing in Bengal in 1908 gave us an angle to examine the international and 'maritime' strategic aims that could be attached to *Swadeshi* activism in Chapter Four. In Chapter Five, he appeared alongside Subramania Bharati as one of the members of the Pondicherry 'Gang', arriving there in 1910, and being something of an outlier as a Bengali in Pondicherry's Tamilian world. Today, Aurobindo is more commonly known as Sri Aurobindo, the spiritual guru whose name adorns the Sri Aurobindo Ashram in Pondicherry and who promoted his own religious philosophy of 'integral yoga'. Yet, when he arrived in Pondicherry on 4 April 1910, Aurobindo was probably the most wanted man in India. He had recently been imprisoned for his part in the Maniktala bomb conspiracy in Calcutta, for which he was later acquitted, and when rumours reached him that he was due to be arrested again, he had escaped to Chandernagore, where he hid for around six weeks before making his way to Pondicherry. Speaking afterwards, Aurobindo said that an *adesh* or spiritual/inner voice had told him to go to Chandernagore, and later, Pondicherry.

Upon arrival in Pondicherry, it became swiftly clear to both the colonial authorities spying on the Pondicherry gang and to the revolutionaries themselves

Geographies of Anticolonialism: Political Networks Across and Beyond South India, c. 1900–1930, First Edition. Andrew Davies.

that far from being a revolutionary figurehead, Aurobindo had decided, based on the spirituality which he had been increasingly influenced by during his time in prison, that he was committed to a life of spiritual retreat and the promotion of his growing religious philosophy. This was as baffling to the revolutionaries as it was to the colonial authorities, who wrote with a degree of suspicion about whether or not Aurobindo had 'withdrawn' from political activity, or whether this was simply a front to stop them arresting him. Over the next four decades until his death in 1950, Aurobindo established a distinct approach to the world which, I argue in this chapter, served to blur the supposed divide between Aurobindo's secular/political work prior to Pondicherry, and his spiritual/inner life work afterwards. This builds on established work in the study of the relationship between colonialism and religion in South Asia which challenges the problematic division between the secular/rational/Enlightenment world and the spiritual/superstitious/precolonial world. In reading of some of Aurobindo's texts and worldviews in this way, this chapter argues that, far from withdrawing from 'the political' as many have seen it, Aurobindo's philosophy was in fact grounded in a universalism which, whilst spiritual, was inherently political too.

In order to make this argument, the chapter first examines the post- and decolonial scholarship on religion which has attempted to destabilise the binary between spiritual/superstition and secular/rational which is often at work in academic studies of religion in postcolonial contexts, as well as exploring how geographical work has intervened with a different emphasis into these debates. The next section discusses Aurobindo's biography to provide some empirical context, before examining his meeting with Mirra Alfassa in 1914 and the effect that this 'cosmopolitan' encounter would have in shaping both of their lives and the city of Pondicherry. A final section elaborates more closely Aurobindo's thought as a potential space of decoloniality or thinking outside the 'Western code' (Mignolo 2011).

Post and Decolonial Religion?

Dipesh Chakrabarty, writing in the final chapter of *Provincialising Europe*, argued that one of the key tensions in writing Indian history was '[scholars] have assumed that for India to function as a nation based on the institutions of science, democracy, citizenship, and social justice, "reason" had to prevail over all that was "irrational" and "superstitious" among its citizens' (Chakrabarty 2000, p. 237). This categorisation, both not only hardened the divide between the 'irrational/superstitious' precolonial and 'rational' colonial eras (and what Chakrabarty calls History I), but also necessarily enrolled various religious or spiritual beliefs into the former. The implicit hierarchy that this creates both not only demeans precolonial beliefs but also cements the supposed rationality of modernity as superior.

Given postcolonialism's (and late Subaltern Studies' in this case) suspicions of the categories of the Enlightenment, it is hardly surprising that Chakrabarty was sceptical of the belief in reason and rationality which to him stifled the heterogeneity of both the epistemological and ontological in South Asia (or indeed any other place where these contradictions play out). To Chakrabarty then, the challenge is not to attempt to translate the religious into the sphere of the rational/secular, but to treat it as its own distinct area, distinct from 'Western' political and social categories. However, this is not unproblematic. To return to Vasant Kaiwar's (2015) critique of Chakrabarty which was discussed in Chapter Three, a problem occurs where, by not 'translating' the religious, we miss out on how religion was an important symbolic component of peasant and workers' rebellions in the colonial era. To Kaiwar, in a Marxist critique of Chakrabarty's Heideggerian project, the point is to engage with these ideas through detailed ethnographies of how people across the world make sense of their situation and to think about how various forms of politics were enrolled in these. Second, Kaiwar's critique points out the tendency in Chakrabarty's writing to reinforce a binary between the 'West' and India – in this case in seeing Western thought and Indian religion as incommensurable. This is a tendency in much of Subaltern Studies that, in rejecting the West, leads to a tendency to create a more pronounced binary between West and India/Orient than is helpful.

Taking on these critiques of Kaiwar's, there remains a need to continue to challenge the supposed clarity of the distinction between secular rationality and religious/spiritual superstition, as is demonstrated through Aurobindo's claims to be guided by an *adesh* in 1910. Sceptical readers could insist that the *adesh* telling Aurobindo to move to Chandernagore/Pondicherry is a post hoc 'spiritualisation' of a moment of very material political duress as the threat of imminent arrest loomed. However, this reading forecloses the fact that, to Aurobindo, listening to the *adesh* and obeying it was a completely *rational* decision, as it meant being guided by his growing spirituality and faith, which was probably equally important to him as any 'worldly' concerns by this point.

Whilst Chakrabarty made his claim some time ago, and *Provincialising Europe* remains one of the most widely cited books written by a member of the Subaltern Studies Collective, it is still striking that very little work on the intersections between post- or decolonial understandings of religion and/or spirituality have emerged. Within the domain of Subaltern Studies Collective, for example, Partha Chatterjee's (1993) work on nationalism famously addressed the encounter between the enlightenment thought and Indian gender roles by 'resolving' them into two different spheres so that modern/enlightened gender relations became coterminous with the public/material sphere, whilst the spiritual aspects became part of an inner/private sphere.

The history of religion in India, and indeed elsewhere, is fraught with tensions about the role of orientalist logics in defining and categorising the various spiritual movements in India – most obviously in the categorisation of Hinduism as a discrete religion rather than a pantheistic system, but which can also be applied to ideas about the encounter between 'Indian' religions and 'Western' thought (Dirks 2001). This tension focusses on the supposed incompatibility of religion with secular liberal democracy and/or multiculturalism and also implies that religion and politics are fundamentally incompatible, or at least where 'political religion' does occur, that it is somehow a corrupted or degenerate form of the supposed rationality of 'the Political'. This framing, which has its roots in the colonial era, but continues today, is what Mandair (2009) calls an 'ontotheological matrix'. Mandair's work here is important as it explicitly sets out to decolonise this onto-theological matrix, and to remind postcolonial scholars that there is still much work to be done in understanding exactly how terms like 'Indian religions' are deeply imbricated with colonial tensions which are still played out today. To Mandair:

> India's passage to modernity happened through its being reawakened to the notion that it once had religion, that it had forgotten its original religion(s), but that this religion could be recovered through the colonizer's benign intervention, in order to progress toward a form of modern national self-governance. As a result, India's passage to globalisation is limited to a stand-off between secularism and/or global religions such as Hinduism, Islam, and Sikhism. Indian society remains at the level of identity politics, at the stage of politics rather than the political, properly speaking, since the latter is not yet achieved. (p. 5)

This is only a part of Mandair's detailed argument but suffice to say for this chapter, that there continues to be a tension between what 'religion' is, and whether it is a suitable term for the variety of religious/spiritual groupings across South Asia, given its colonial legacy.

Here it is also worth mentioning that this is further complicated by the inter-sections between the religious/spiritual resurgence in India in the fin-de-siècle (discussed in Chapter Three), anticolonialism and the freedom struggle, and, the enlightenment thought of 'The West/the Coloniser'. Thus, the simplistic opposi-tion of secular/rational with religious/irrational should be problematized. Taking Mandair's point, and in a Tamil context, Ravi Vaithees (2015) has argued for a more processual account of 'religion-making' (Mandair's term) as it exposes how 'the intellectual and cultural foundations for a diverse range of often contending subjectivities, ethno-nationalist, and identitarian movements were formed at the same time as the emergence and consolidation of a pan-Indian Hindu nationalist ideologicial/discursive formation' (p. 11). For Vaithees, the work of Tamil reli-gious reformer Maraimalai Adigal during the nineteenth century 'Neo-Saivite' Movement was essential in opening up space for the later anti-Brahmin political movements that have dominated (Indian) Tamil politics and nationalism ever since. Whilst a debate about Indology is beyond the scope of this book, what is important here is that this work has begun to show that, as was discussed in Chapter Two, the concept of 'the political' is often still rooted through a Schmittian understanding of definite political categories, in this case of the opposition of secular and spiritual. Further, as noted in the discussion of political violence in India and how radicals like Bal Gangadhar Tilak appropriated texts like the Bhagavad Gita to make the case for the centrality of violence to Indian (South Asian) political life which were discussed in Chapter Three, as well as Subramania Bharati's *bhakti*-inflected writing in Madras Presidency show, the supposed boundaries between spiritual and secular politics were always porous. A more nuanced reading of the intersections between religion and the political is therefore necessary to understand the history of 'Indian religions', and this continues to have important effects into the present.

Similar trends in interrogating the spaces of religion/theology can be found in the huge amount of work on religion across social and cultural geography since the cultural turn of the 1980s (Dwyer 2016). Holloway (2011), for example, has argued that geography needs to take seriously the theological sensibilities of the individuals and groups who are the subjects of our studies. To Holloway, traditionally, geographical approaches to religion have been either 'substantial' or 'situational'. The substantial is more phenomenological and seeks to understand what the essential nature or experience of the sacred is for those who believe. As a result, it tends to downplay the very situated and contested ways in which concepts and experiences like 'sanctity' have been produced and as a result does not critically interrogate these concepts and their making. The substantial approach then sees the research as taking place 'inside' the subject of study.

The situational approach, in contrast, attempts to uncover the social, political, economic (and other social categories) processes and practices which shape the sacred. Thus, instead of being 'inside', situational studies of religion seek to develop a sense of critical distance – or attempt to be 'outside' – which has the problematic effect of rendering the researcher as somehow objective and impartial in their academic judgement of how religion, theology or belief is experienced by those who are studied. This judgement is also often framed in sociological terms of economy/politics, etc., which are not necessarily compatible with religion or faith. This has obvious similarities to the debates about the postcolonial binary of rationality and superstition which was discussed earlier. To Holloway then, the challenge is to conduct research that fits 'between' these two approaches, and one way of doing this is to utilise affective and more-than-representational geographies (Lorimer 2005; Anderson 2006) to understand spaces as immanent assemblages, where the rhythms, materialities and sounds intersect to produce and compose religious spaces.

However, whilst the likes of Hill (2015) have explored how the non-representational geographies of the past can be understood through the experience of their traces in the present, my approach is not to write ethnographically or more-than-representationally about the past and present spaces of the Sri Aurobindo Ashram, or Auroville, the experimental city north of Pondicherry started by a group of Ashramites in the 1960s. This is because, rather than the study of religious space *per se*, the point of this chapter is to examine how the life and work of Aurobindo can be read through a decolonial lens as an attempt at a form of cosmopolitan pluriversality. Whilst Walter Mignolo has recently attempted to draw together some of the tensions between the 'Western' phenomenology of Husserl (which shares some similarities with more-than-representational geographies) and the decolonial (Mignolo 2018), here I read Aurobindo through the lenses provided by recent work on the cosmopolitan intellectualism and anticolonialism. As noted in previous chapters, the turn towards global histories and historical geographies has exposed the complex interconnections which occurred within colonial and imperial spaces (Raza et al. 2015). Manjapra (2010) has argued this emphasis has made it clear that viewing these geographies as simplistic cores and peripheries only emphasises the imperialist view of the world and its attempts to divide the world according to the criteria which imperialism/colonialism valued. Thus, rather than seeing a universalistic 'cosmopolitanism' rooted in Western modernity, we should recognise that there were always a range of cosmopolitan-isms at work which were drawn from the individuals concerned and their own backgrounds, as well as the spaces in which any cosmopolitan encounter occurred. As Holton (2002) has argued, this has two important consequences, first that rootedness within a particular place or cultural context does not limit the ability to identify as a citizen of the world or as part of a wider humanity, and second that cosmopolitanisms were not something that citizens of the global south 'learned' from their supposedly more 'modern' comrades. Elsewhere, Mukherjee (2018) has shown how these cosmopolitanisms intersected with gender as well – the interaction of feminist 'Western' activists like Annie Besant in the Theosophical Society was often seen as more acceptable to some Indians because of her spiritual positioning in the Society. However, these were placed alongside other racialised understandings by women like Margaret Cousins who saw Indians as better able to communicate with other Asian women, but also that Indian spirituality could act as a counter to Western modernity (Candy 1994).

The intersections between cosmopolitanism and faith have proved a useful way of understanding how commonality was bridged across difference during this period. Leela Gandhi's (2006) work on friendship is foundational here, and indeed, it is notable that she draws upon the cosmopolitan encounter between Sri Aurobindo and his consort, Mirra Alfassa (known to their followers as The Mother), as a key space in which anticolonial communities were forged – and more on this later. However, more recently, Haggis et al's. (2017) edited collection has provided a range of examples of the ways in which faith provided a vehicle by which the varieties of difference produced in the colonial world could find grounds for what Gandhi (drawing on Jean-Luc Nancy) would call compearance – the being in common of singularity.

However, 'cosmopolitanism' is also a word freighted with elitist connotations – the traditional idea of the 'citizen of the world' is often framed as some sort of dilettante individual who roams the world freely, sampling what they want (Manjapra 2010). In this, there is, as Menon (2015) states, a danger that the turn towards writing intellectual and boundary-crossing (in both the geographical and ideological sense) cosmopolitanisms that is taking place at present across a range of literatures could potentially be read as an elitist counter to the subaltern – emphasising the often elite spaces by which cosmopolitan intellectuals encountered each other and their mutual ideas. For example, it could be thought that the four men at the heart of this book, as members of high-caste groups who were all well-educated and active in the production of knowledge, represent elitist and nationalist histories which prefigure the bourgeois nationalism of Indian independence. However, this occludes the range and scope of activities which made up the lives of these men – all were involved in work with groups which are non-elite (for example Pillai's labour organising in Tuticorin, or Bharati's attempts to combat caste and gender discrimination). As a result, reading intellectual cosmopolitanism only as a form of elitism is deeply problematic, as it both presupposes that non-elites cannot possess or engage with intellectual or cosmopolitan values, but also that the boundaries between elite/non-elite/subaltern are somehow fixed or clear.

Of especial use here is the variety of work done by geographers. Gidwani's work on the subaltern cosmopolitanisms of the present has emphasised the ways in which those who are notionally 'subaltern' construct complex ethical and moral logics about global capitalism and climate change (Gidwani 2006). Elsewhere, David Featherstone has consistently challenged the supposed inability of subaltern groups to be cosmopolitan and crucially has shown how these cosmopolitanisms were spatialised (Featherstone 2009, 2012, 2013, 2019). As was noted in the discussion of anticolonialism in Chapter Two, as much as people like Gandhi were engaged in thinking and writing across Western, non-Western, and potentially hybrid intellectual registers, so was Aurobindo – or Bharati, or any other writer in this book. As Bose (2007) has forcefully argued, this is especially necessary for Aurobindo as he has been variously misappropriated or misread – either as a harbinger of communalism or as lacking the humanism of Gandhi (on this, see also Heehs 2006). The key point of this chapter then is to think through exactly what this cosmopolitan approach can do to ensure that we recognise and treat fairly the politico-religious thought of individuals like Aurobindo.

The Cosmopolitan Politics and Spirituality of Aurobindo's Life

Aurobindo was only active in formal, nationalist politics for a few years, yet his writing and work has proved to be important in establishing a number of political frameworks through which anticolonialists in India could make sense of their struggle. As an avowed member of the extremist wing of the Indian National Congress (INC), Aurobindo wrote extensively about the need for radical resistance to British rule, but was firmly positioned as an intellectual figurehead rather than a revolutionary of the street/barricades.

Born in Calcutta in 1872, Aurobindo's father was an Anglophile who in 1879 sent Aurobindo and his two brothers to England to be educated in the hope that they would gain entrance to the elite Indian Civil Service (ICS). Of the brothers, Aurobindo was the most intellectually suited to the ICS, gaining entrance to King's College on a scholarship in 1890 to study for entry. However, despite his father's wishes, Aurobindo had little interest in the ICS, and deliberately disqualified himself by arriving late for a horse-riding examination. Aurobindo managed to secure a job working in the Princely State of Baroda, returning to India in February 1893. He worked in a number of state departments, before becoming a teacher and later Vice-Principal of Baroda College in 1897.

Gradually, Aurobindo became more influenced by the growing Indian nationalist movement, and began writing for various publications and established a number of contacts within extremist circles. Aurobindo's brother Barin was also heavily involved in nationalist politics. In 1906, following the Partition of Bengal, Aurobindo moved back to Calcutta and was involved in the 1906 Congress meeting at Calcutta (with Dadabhai Naoroji as President), and the bitter split between the Extremists and Moderates at the 1907 Congress in Surat.

Aurobindo's writing at this time was foundational in thinking through what freedom meant in an Indian context, as well as setting out a number of foundational tenets of the concept passive resistance. As Mahajan (2013) points out, Aurobindo was at the forefront of *Swadeshi* writing which made the case that ancient Indian civilisation was different to 'Western' civilisation, and that this difference was rooted in *Vedanta* – as Aurobindo stated in 1908, 'the final fulfilment of the Vedantic ideal in politics, this is the true *Swaraj* for India' (Ghose 2002, p. 1086). *Vedanta* is one of the schools of orthodox Indian 'Hindu' philosophy and is an umbrella term for a range of traditions, but which are all interested in the relationship between the *Brahman*, or metaphysical reality, the *Atman* or individual soul/self and the *Prakriti*, or material world/reality. A detailed discussion of the various schools of thought and specifics of Vedantic ideals is unnecessary, but Aurobindo's positioning here had two important consequences.

First, it shows how Aurobindo was at the forefront of the Hindu revivalism which was taking place and which intersected with politics in the spiritual/political ideal of *Swaraj* discussed in Chapter Two. Second, this framing has had long-lasting consequences as it has meant that the concept of freedom in India is closely associated with *Swaraj*, which in turn is deeply imbricated with Hindu, and therefore majoritarian politics. As Heehs (2006) has pointed out, this has meant that Aurobindo's political thought is often mobilised as a revisionist precursor to today's Hindu-right majoritarianism, and he is either lauded or vilified as a result. This also leads to the problematic readings of Aurobindo's writing. An example here is the pamphlet *Bhawani Mandir* (The temple – *Mandir* – of the goddess *Bhawani* – who is an incarnation of the goddess *Kali*), written in 1905 by

Aurobindo in Baroda. Famously, the Rowlatt Sedition Committee Report drew upon Criminal Investigation Department (CID) readings of the pamphlet to declare that it was 'a remarkable instance of the perversion of religious ideals to political purposes' (Rowlatt Committee 1918, p. 24) and that it was one of three 'mischievous or specially inflammatory' books which were in circulation at the time. This, as Heehs discusses, had important consequences on the future inter-pretation of Aurobindo, as the pamphlet itself was not especially revolutionary nor political, but rather it was written for a social-service organisation. It was also Aurobindo's brother Barin's idea to write the pamphlet, which was soon forgotten by the revolutionaries themselves. This highlights the problematic nature of reading the documents of the colonial authorities, willing as they were to see evi-dence of political agitation in the most minor of places, as well as the well-known issues with the Rowlatt Report and its subsequent Act (which in turn sparked a massive wave of agitation in 1919 and indirectly led to the Jallianwala Bagh mas-sacre in Amritsar in 1919 – see Wagner (2019)).

I will explore more of Aurobindo's 'political' writing later, but suffice to say that he wrote extensively during his involvement with the *Swadeshi* movement. This 'political' writing most often in the English language not only in *Bande Mataram* but also in a variety of other newspapers, pamphlets and other sources, such as the Bengali *Jugantar*. As Wolfers (2016) points out the political activities which Aurobindo was involved in at this time took place outside the existing political structures of the INC, and it was the dynamic and activist nature of what was occurring in Bengal which inspired Aurobindo. This mainly involved the development of self-organised societies known as *samitis* which had been devel-oping since the later years of the nineteenth century. Aurobindo had been involved in organising *samitis* since 1902, and in 1906 organised with his brother Barin a *Jugantar* Party along these lines. These *samitis* were intended to become the core of a guerrilla movement with which to bring down the colonial state over the course of decades. However, this long-term plan did not fit with the growing desire for immediate action amongst the radicals in Bengal. Thus, political vio-lence, and specifically assassination attempts through bombs/grenades were attempted. Whilst Aurobindo was sceptical of the capacity of a small number of bombings to challenge the colonial state, he did not stop them from taking place (Heehs 2008).

During this time, it was Aurobindo's brother Barin who took the lead in organ-ising and plotting revolutionary activity, whilst Aurobindo was focussed upon writing and journalism. Amongst the wealth of daily reporting, Aurobindo wrote a series of seven articles in *Bande Mataram* between 11 and 23 April 1907 enti-tled *The New Thought: The Doctrine of Passive Resistance* (Ghose 2002). These set out the methods and limits to the emerging tactic of the *Swadeshi* movement, but provided the extremists with a clear framework by which to argue against and resist both colonial rule and the more reformist position of the moderates. Here it is worth remembering Hardiman's (2013) critique of the Subaltern Studies

Collective's treatment of 'passive resistance' as a form of bourgeois counter-revolution discussed in Chapter Three – Aurobindo's 'passive resistance' was definitely to be considered alongside political violence as a potential tool for anticolonial struggle, and again, we should resist the temptation to think that passive resistance, in this case, was somehow an opposite to violence.

Given the rising violence in Bengal and elsewhere in India, it was unsurprising that a crackdown on *Swadeshi* activism occurred. In May 1907, Lala Lajpat Rai (one of the Lal, Bal, Pal triumvirate of extremists) was deported alongside the fellow revolutionary Ajit Singh to Mandalay. In a series of articles in *Bande Mataram* that month, Aurobindo denounced these countermeasures, arguing that they were making clear British despotism and were removing the façade of benevolent rule. Wolfers characterises Aurobindo's approach here as a form of revolutionary asceticism, where the brutality of the colonial state would harden the resolve of those who were struggling against it, and that suffering was an essential part of this, and this is 'a broadly Vedantic form of Hegelian idealism' (Wolfers 2016, p. 529).

When the state tried to prosecute Aurobindo in August 1907 for the publication of seditious articles in *Banda Mataram*, they called upon the former editor and fellow extremist Bipin Chandra Pal to testify against him. Unsurprisingly, he refused, and was jailed for six months. It was Pal's release from prison that acted as one spark for the agitations in Tinnevelly alongside the SSNCo. However, the crackdown on *Swadeshi* organising, meant that the *Jugantar* Party and other *samitis* were driven underground (Sarkar 2010 [1973]). This did not stop the groups from plotting. The *Jugantar* Party, by this time also known as the 'Maniktala' or 'Maniktala Garden' Secret Society from the location of their 'ashram' in Calcutta, had around a dozen young men willing to be trained in revolutionary violence under the instruction of Aurobindo's brother Barin and their fellow revolutionary Hem Chandra Das, who had been to study bomb-making in Paris in 1907. The attempts by the Maniktala society to commit acts of violence were limited – the 10 attempts at assassination, train derailings or 'dacoity' (violent armed robbery) all failed. The most dramatic failure in April 1908 was when two of the group, Khudiram Bose and Prafulla Chaki, killed the wife and daughter of Pringle Kennedy a British Barrister in the town of Muzaffarpur. The two men had been attempting to kill the Magistrate Douglas Kingsford, who had presided over Aurobindo's earlier trial and who had been targeted by the group before, but when confronted with two identical carriages, Khudiram and Prafulla had thrown a bomb at the wrong one.

Within 36 hours of the deaths of Mrs. and Miss Kennedy,[1] all members of the Maniktala society were rounded up, although Chaki committed suicide before arrest. Khudiram was sentenced to death and hung in August 1908 at the age of

---

[1] The given names of these two women are not noted in the colonial archive.

18. The case of *The Emperor vs Aurobindo Ghose and Others*, more commonly known as the 'Alipore Bomb Case' (named after Alipore Jail where the group were confined), started on 18 May and stretched on for six months. By coincidence, Charles Beachcroft, the sitting judge, had passed the ICS entrance exam with Aurobindo in England. The long hours of solitude in jail provided space for meditation and spiritual retreat. Aurobindo began to be guided by the voices he heard during his seclusion, and an *adesh* telling him that he would soon be released to do great work for India. This work would be spiritual in nature, as God was giving India freedom for the benefit of the world (Heehs 2008). Whilst Aurobindo experimented and developed his spiritual agenda, the trial's focus lay on proving his complicity to the activities of his brother and revolutionary associates, and reams of evidence was produced. It was not until 6 May 1909 that the court was ready to deliver its judgement, and whilst the majority of the members of the society were found guilty, including Barin, Beachcroft found that there was not enough evidence to prosecute Aurobindo, who was acquitted.

The results of the trial were astounding news to many, and word of this judgement spread rapidly. In Pondicherry, Bharati celebrated the news by publishing the *India* with a front page cartoon of 'The Rahu[2] which came along to swallow Aurobindo's sun is slinking away', with the image of a foreign snake-headed god being driven away by the light of the sun, which features a stylised image of Aurobindo in its centre (*India*, May 1909, Sri Aurobindo Ashram Archives).

Upon his release, Aurobindo started a new paper, the *Karmayogin*, but made few public appearances, and the tone of his writing was less strident that before his arrest, and eventually he stopped writing about formal politics altogether. He was still a target for suspicion from the colonial state, and when in February 1910 he was warned that he was due to be arrested, he left Calcutta having heard an *adesh* telling him to go to Chandernagore. Hiding out for a month in the French *comptoir* north of Calcutta, he eventually secured travel on the SS *Dupleix*, a French-owned steamer named after the most successful governor of French India in the eighteenth century, and arrived in Pondicherry on 10 April 1910.

The members of the Pondicherry 'gang' were undoubtedly excited to have Aurobindo in their midst, as to them, he would provide a means for the spreading of the message of *Swaraj* from Pondicherry across Madras Presidency and beyond. However, from the beginning it was clear that Aurobindo wished to use the solitude of Pondicherry to develop his spiritual methods, as his practice of yoga was becoming increasingly all-consuming. He was rarely seen outside his residences in Pondicherry, something which caused a degree of suspicion that he was still involved in nationalist plotting by the Madras CID (GO No. 1335, TNA). However, Aurobindo's 'retreat' into spirituality was exactly what it seemed, a

---

[2] A Rahu is an Indian astrological symbol in which is associated with inauspicious events such as eclipses and is often represented as a snake.

move away from nationalist politics towards engagement with his 'inner' world. To most then, Aurobindo's life falls into two distinct phases, pre-1910 his material and political existence, with its most 'political' phase occurring between 1906 and 1908, followed by a 'spiritual' phase from 1910 where he withdrew from public life and focused on his spiritual life's work until his death in 1950. However, as I wish to explore in the rest of this chapter, the clear-cut divide here is deeply problematic. The next two sections of the chapter explore first the cosmopolitan encounter between Aurobindo and Mirra Alfassa in Pondicherry, and then the ways in which thinking decolonially helps to further challenge the boundary between the political/spiritual binary in Aurobindo's life.

Cosmopolitan Encounter with the Mother

Walk into any enterprise in Pondicherry today, and it is likely that there will be two photographic portraits on display. One is of Sri Aurobindo. The other depicts The Mother, Aurobindo's close collaborator, who was born as Mirra Alfassa in Paris in 1878. The connection that was established between the two shaped the geography of Pondicherry, establishing it as a centre for yoga and spirituality, but is also crucial to understand some of the cosmopolitan intersections between individuals that took place amongst utopian forms of radicalism in the early twentieth century.

Alfassa first visited Pondicherry in 1914 with her then-husband Paul Richard, who was seeking election to the French Senate through Pondicherry's constituency. Both Alfassa and Richard were travellers through the varied cosmopolitan utopianisms of the fin-de-siècle, and met through Max Theon, the founder of the Cosmic Movement, an occultist organisation based in Tlemcen in Algeria. Richard had visited Pondicherry before and had met Aurobindo in 1910, but in 1914, both Mirra and Paul Richard met with Aurobindo, and this encounter had profound consequences. Mirra recognised Aurobindo as the individual who had been appearing to her in visions, and together the two Richard's planned with Aurobindo the publication of his ideas in a new philosophical journal, the *Arya*. These publications formed the basis of Aurobindo's expanding philosophical thought system, which came to be known as 'integral yoga', and were later published as *The Life Divine, The Human Cycle, The Synthesis of Yoga*, and more (Mohanty 2015).

Due to their association with Aurobindo, the Richard's were ejected from Pondicherry in 1915, but Mirra returned to Pondicherry in 1920, and shortly after separated from Richard. Alfassa's encounter with Aurobindo was so profound that she abandoned her other spiritual and occultist paths and became Aurobindo's spiritual consort, who Aurobindo regarded as his equal. This meeting of anglicised Bengali revolutionary-cum-guru and French occultist and bohemian in the creation of a new spirituality is nothing if not cosmopolitan but also tells us something else if we read it decolonially. The encounter between

Aurobindo and Alfassa shares similarities with the spiritual anticolonial deep relation of Robbie Shilliam (2015) discussed in Chapter Two. To Shilliam, the spiritual remains a relatively uncolonised space, providing an arena where two different groups, in his case black and Polynesian activists/anticolonialists, can connect with each other and learn about their respective singularities – their 'I and I' (Shilliam 2015, p. 31) – through thinking about how they envision the world spiritually. This is more subtle than saying that the spiritual is some uncontaminated or autonomous 'zone' free from colonial or imperial interference, but rather to say that the spiritual is often the place where these influences are less likely to have such profound effects upon the colonised. This is useful in thinking about Aurobindo and Alfassa as it allows us to move their encounter away from colonised categories such as the stereotype of 'East meets West'. Instead, we should recognise that it was through their spirituality that allowed them to transcend the boundaries of class and race which the colonial world had created, and which should have kept them apart as radically 'other' subjects of empire.

The meeting between the two was life-changing for both of them. Shortly after her return to Pondicherry in 1920, Aurobindo began calling Alfassa 'The Mother', a practice that the other inmates of Aurobindo's house followed. Again, gender is an important factor here – the epithet of 'The Mother' again draws parallels to the maternal characteristics which are often applied to women in Indian society – such as *Bharat Mata* (Ramaswamy 2010). Given that Aurobindo was one of the prime proponents of the idea of *Bharat Mata* in his explicitly political writing, his decision to name Alfassa is significant. It also indicates some of the ways which, as noted above, gendered norms continued to play a significant role, even in the cosmopolitan spiritualist circles which the Richards lived in, and which Aurobindo had begun encountering when he moved to Pondicherry.

As The Mother, Alfassa was integral in the establishment of the Sri Aurobindo Ashram, and as Aurobindo became more and more secluded in his yogic practises, it was Alfassa who organised the day-to-day running of the Ashram. The formal date often given for the establishment of the Ashram is 24 November 1926, the day that Aurobindo withdrew from public view and handed over the running of the Ashram to Alfassa. However, far from being simply an administrator, Alfassa was also working alongside Aurobindo in the development of integral yoga – based on the belief that, unlike previous ascetic yogic practises, it should be practised in everyday life.

Alfassa's stewardship meant that, after Aurobindo's death in 1950, she became the figurehead of the Ashram. After a period of withdrawal in the 1950s following Aurobindo's death, Alfassa began a project to create the intentional community of Auroville – a 'universal city in the making', which was eventually formed on land about 10 km north of Pondicherry, and thus in the state of Tamil Nadu, in 1968 according to a design by Roger Anger. Visitors to Auroville today are required to listen to a recording of The Mother's words explaining the philosophy of the city as a space to experiment with human unity. These are set out in a four-point charter written by Alfassa (1968) for use in the city's inauguration ceremony:

1. Auroville belongs to nobody in particular. Auroville belongs to humanity as a whole. But to live in Auroville, one must be the willing servitor of the Divine Consciousness.
2. Auroville will be the place of an unending education, of constant progress, and a youth that never ages.
3. Auroville wants to be the bridge between the past and the future. Taking advantage of all discoveries from without and from within, Auroville will boldly spring towards future realisations.
4. Auroville will be a site of material and spiritual researches for a living embodiment of an actual Human Unity.

Whilst it was envisioned as a city of 50,000 people, today the population is around 2500, from c. 49 countries, but mostly of South Asian origin. At the centre of the city is the ceremonial space of the 'peace zone', which contains an inaugural urn with the soil from 124 nations and 23 Indian States, as well as a paper copy of the charter handwritten by Alfassa in French deposited in it, alongside the giant, gold-plated Matrimandir, the most notable 'landmark' of Auroville today – a meditation chamber which contains the largest perfectly spheroid glass orb in the world to reflect light around the chamber. Despite its utopian intentions, as a city, Auroville is subject to the some intense contestations and tensions (Namakkal 2012; Jazeel 2015) – siting a huge gold-plated mediation chamber in the middle of landscape that was (and still is to their residents) a group of Tamil villages, who are often employed as wage labourers by Aurovillians, is deeply problematic. Whilst I was writing this chapter, Jessica Namakkal pointed out these tensions and their neocolonial nature in a Twitter thread challenging the utopian reporting of Auroville as an alternative to polluted urban India by Buzzfeed India (see https://twitter.com/j_namakkal/status/1064900916053196801).

The Sri Aurobindo Ashram and Auroville, whilst separate entities administratively, are both closely connected to this Aurobindonian tradition, and both have a significant presence in and around the city of Pondicherry (the cover image of this book is, for example, an image of an Ashram Building in Pondicherry's 'Heritage Quarter', and the Ashram is one of the largest landholders in the city today). The evolution of the Ashram and city of Auroville over the past century is thus both striking and often problematic. However, the confluence of cosmopolitan and utopian imaginaries which the encounter between Aurobindo and Alfassa created also goes to show that the intersections of anticolonialism, spiritualism, radicalism and utopian imaginaries was productive of wildly heterodox spaces. As Leela Gandhi puts it in her chapter discussing Alfassa in *Affective Communities*:

[It is possible to] safely attribute the startling coincidence of mystical and revolutionary impulses in Alfassa's practice to the ideological cartography of the fin-de-siècle spiritualism in which she found herself implicated. In that set of historical conjunctures the path to heterodox theism, as with the alternative movements of our own time, often passed through the sidestreets of political radicalism. (Gandhi 2006, pp. 124–125)

The encounter between Alfassa and Aurobindo, then, is indeed 'startling' both not only in the longevity of the relationship between the two but also in the radically alternative spaces which were produced. I imagine it would have been difficult for Charles Beachcroft, the judge in the Alipore Conspiracy trial, to countenance that his old college-mate would soon end up being held up as the divine avatar of a new form of yoga, and 50 years later would see his ideas inspire the birth of a new city, however problematic that city is today. The final section of this chapter will examine how thinking decolonially can help to make sense of Aurobindo's position, and in doing so, can work through some of the supposed binaries between rationality/modernity/secularism and irrationality/premodernity/spiritualism.

Aurobindo's Thought and Practice as 'Decolonial' Challenge to the Political

As the discussion in this and previous chapters has made clear, the building of Indian nationalism and the freedom struggle was closely related to the spiritual resurgence of India in the latter half of the nineteenth century. Aurobindo was not alone in articulating the belief that India's freedom was predicated upon the re-establishment of Indian philosophical and spiritual categories. As discussed earlier, to some, that these were based upon 'Hindu' philosophies positions Aurobindo and others like Vivekananda as precursors to the right-wing Hindu majoritarian politics at work in India in the late twentieth and early twenty-first centuries (Heehs 2006; Wolfers 2016). However, following Zachariah (2015), this presupposes that the Hindu revivalism of the fin-de-siècle is the same in scope and aims as the politics of Hindutva today. Aurobindo's removal from 'formal' political activity in 1910 meant that he did not explicitly speak about these concerns in his later life, which is different to other thinkers, such as Mohandas Gandhi, who later moderated or altered their positions with regards to Indian nationalism's connections to 'Hindu' social values, like the caste system (although Gandhi in particular never resolved the contradictions which the Dalit-Bahujan leader B.R. Ambedkar challenged him upon).

It is in the writing of (and about) Aurobindo that we can find a certain degree by which he was himself grappling with the binaries by which his life was determined by others as either/or spiritual/political at various stages. This is particularly notable for Aurobindo, given that he is not only an important figure in Indian history and culture who has had a huge amount written about him but also due to the recent controversy surrounding the publication of Peter Heehs' (2008) biography *The Lives of Sri Aurobindo*.

Widely recognised as the most detailed and scholarly work on Aurobindo upon its release, Heehs' book was based on a lifetime of research and of being a member of the Aurobindonian tradition – he was a researcher in the Sri Aurobindo Ashram Archives in Pondicherry at the time. Upon its release, a group of Ashram members objected to Heehs' attempts to render Aurobindo's experiences as 'human', as to them he was an avatar of the divine. Without wanting to go into details about the whole controversy when it is still available in the public domain (and is summarised in Prince 2017), one of the core issues within the controversy was the question about whether and how it is possible to 'objectively' understand Aurobindo's faith (including the visions and voices which guided him) from a secular perspective. Heehs' attempted to strip away some of the supposed mysticism of Aurobindo – including choosing a non-airbrushed image of him for the front cover of the book – in an attempt to create a more objective study. The controversy had serious consequences, with public pressure on Heehs meaning the book's publication was delayed in India, his roles within the Ashram were removed from him, and that his Indian visa was nearly not renewed. In response, Heehs supporters argued that the work was one of scholarly rigour which should be supported, and that his critics were too narrow in their conception of how to interpret Aurobindo's life.

On the one hand, such debates are often held up as symptomatic of an increasingly intolerant strand in Indian society which attacks authors who do not write histories of religions which conform to prevailing orthodoxies – a similar but much more heated controversy occurred around the publication of Wendy Doniger's *The Hindus: An Alternative History* in 2009. However, as the section on religion and the post/decolonial discussed earlier, the debate about the biography of Aurobindo highlighted the continuing tension between the secular/rational. To find a way through this, my interest here is not in the theological debate *per se*, nor in a detailed discussion of Aurobindo's Vedantic principles – although see Prince's (2017) book which argues for a detailed reading of these, and which hints that the lack of scholarly engagement in Aurobindo lies in the incommensurability of Aurobindo's thought within disciplinary boundaries.

Instead, my argument here lies in the field of decoloniality, and the ability to read Aurobindo as a pluriversal thinker whose very being challenged colonial categorisations. In this, I am following Walter Mignolo's engagement with Anibal Quijano and the concept of coloniality/modernity. As discussed in the introduction, these Latin American ideas are important as they provoke a consideration of the epistemological and ontological categories within which academia functions and how they constantly reinscribe colonialist assumptions on our theory and practice. In particular, it is useful to think about Aurobindo through Mignolo's argument in *The Darker Side of Western Modernity* (2011, p. xii) that the world is dominated by a 'Western code', or 'the belief in one sustainable system of knowledge, cast first in theological terms and later in secular philosophy and sciences … [which] is pernicious to the well-being of the human species and to the life of the planet'. Reading Aurobindo in this way is also likely to be counter to the readings of those Ashramites who objected to Heehs' book, but I think it is worthwhile as it continues to push at and destabilise this 'Western code'. This is different to claiming that Aurobindo's dedication towards seeing his spirituality as being of benefit to all of humanity, and therefore being of social value (Ambirajan 1995a, 1995b), although this is of course true to some extent. Instead, it is about worrying away at the spaces outside the Western code which people like Aurobindo created, and exploring how these intellectual spaces were understood. Here it is important to recognise, as Heehs does (2008, pp. 189, 211), that Aurobindo was essentialist in his universalism, and also in his belief that Indian culture was superior to other nations. In this, he was undoubtedly a product of his time and the utopian circles which he moved within. It is also true that Aurobindo, like V.O.C. Pillai and Subramania Bharati, was happy to be seen as a nationalist, but we must again be careful in reading the nationalism of early twentieth-century India as a simple precursor to the nationalism we understand today, whether in India or elsewhere. Whilst recognising these limits, instead the next few pages begin to think about both not only how the authorities failed to make sense of Aurobindo's supposed transition from politics to spirituality but also how he himself made sense of this positioning.

The colonial 'Western code' is clearly visible at work in many of the colonial records of Aurobindo's life. As the Rowlatt Report's (mis)reading of Aurobindo's *Bhawani Mandir* discussed above showed, even in 1918, they saw his spiritual writing as nothing more than an attempt to 'twist' religion to suit political ends, rather than the more straightforward piece of social work that the pamphlet was. However, once in Pondicherry and fully immersed in the so-called 'spiritual' aspects of his life, the inability of the colonial authorities to categorise Aurobindo becomes clearer. Reading through the confidential files on Aurobindo that were kept by both the Madras CID and other colonial figures like the British consul in Pondicherry, it is clear that Aurobindo's withdrawal from politics was completely baffling to the authorities. Superintendent Longford's report on the 'Anarchists' in Pondicherry (GO 1335, 1911, TNA) details the seclusion of Aurobindo, but continues to be suspicious of him as a key plotter, so that, even in the straightened circumstances of the revolutionaries in August 1912, he believed that

Arabindo Ghose [*sic*] is important as he commands general respect … [He] says in reply to a friend that he will not go back to Bengal as there will be trouble there shortly and the police will be sure to attribute it to him if he is back; … I can see no connection between [the 'gang'] and any centres in Madras City. But I do not know how far the wires are pulled from Bengal through Arabindo Ghose [*sic*] and his party. But they will certainly require a larger subsidy if they are to be really active.

That Aurobindo was happy in spiritual seclusion and had removed himself from the visible political struggle is still unbelievable to Longford, over two years after Aurobindo's arrival in Pondicherry. The colonial authorities' suspicions of the 'anarchists' in Pondicherry was understandably difficult to remove, especially given the role of the Pondicherry 'Gang' in the assassination of Ashe in 1911, and the close spatial proximity of Aurobindo meant that he could not be cleanly absolved. This is despite the fact that most records simply show that Aurobindo kept to himself and very rarely received guests. Some of his close Bengali associates did engage with different groups around the city, and thus gave room for the more paranoid elements of British imperial surveillance to keep him under surveillance. However, Pondicherry's decline as a centre for anticolonial organising, as well as Aurobindo's willingness to remain there – as noted by the Deputy Inspector General of Police in Madras in a letter to the Consul in Pondicherry in 1914 (Diary of His Britannic Majesty's Consul at Pondicherry, NAI) – meant that it was increasingly clear that Aurobindo had given up 'Politics' as the Government of India (GoI) and the CID in Madras saw it. His spiritualism therefore meant that he was shifted into a different, less-threatening, category of colonial subject, who would still be observed from time to time, but was not an imminent threat to the colonial order, as he was seen to be in Bengal. To the authorities then, this was something of a lateral move, shifting Aurobindo and his followers from one 'box' in the colonial order of things to another, whilst keeping him firmly placed as a subject of empire, albeit a seemingly irrational one.

However, Aurobindo himself resisted the binary between Politics/Spirituality as distinct categories. Writing in 1920 to Joseph Baptista, a barrister for Bal Gangadhar Tilak's Socialist Democratic Party who had written to Aurobindo asking him to take up the editorship of new newspaper in Bombay, Aurobindo, whilst rejecting the offer, clearly stated an objection to the supposed divide between the spiritual and secular:

> I do not at all look down on politics or political action or consider I have got above them. I have always laid a dominant stress and now I lay an entire stress on the spiritual life, but my idea of spirituality has nothing to do with ascetic withdrawal or contempt or disgust of secular things. There is to me nothing secular, all human activity is for me a thing to be included in a complete and spiritual life, and the importance of politics at the present time is very great. But my line and intention of political activity would differ very considerably from anything now current in the field. (Ghose 1995, p. 142)

In this, Aurobindo is more nuanced than in many of his writings, as he was often quite blunt in turning away the many requests he received for his thoughts about current political events taking place in India, and often did not challenge the secular/spiritual divide as clearly as he does here. But it is clear here that he was thinking about the wider importance of his spiritual life.

Aurobindo also believed that his utopian and universalist convictions about integral yoga were to be of benefit to all of humanity, and far exceeded in importance something like the independence of India, although that was a useful stepping stone along the way. In his Independence Day message in 1947, where he explicitly argued that the independence of India was only a stepping stone towards a future of pan-Asian unity, to be followed by an eventual world union, and an evolutionary change in consciousness (Ghose 1995). This position is made clearer in his earlier writing. For instance, in *The Renaissance in India*, which was collectively written in segments between 1918 and 1921, but here in one of the earliest sections composed in 1918, Aurobindo wrote:

> The method of the West is to exaggerate life and to call down as much – or as little – as may be of the higher powers to stimulate and embellish life. But the method of India is on the contrary to discover the spirit within and the higher hidden intensities of the superior powers and to dominate life in one way or another so as to make it responsive to and expressive of the spirit and in that way increase the power of life. … The work of the renaissance in India must be to make this spirit, this higher view of life, this sense of deeper potentiality once more a creative, perhaps a dominant power in the world. (Ghose 1997, pp. 15–16)

Here we can see that the 'spiritual' Aurobindo is convinced of the pre-eminence of Indian philosophy over that of 'the West', but crucially that this is an engaged worldview that seeks to change to world and not to retreat from it. This spirituality is universalist and utopian in orientation, and the use of the word 'dominate' by an anticolonial thinker shows that Aurobindo was not a 'pluriversal' thinker in the way that Mignolo would suggest – where the world should be recognised as a epistemically diverse. However, his worldview and his vision of a spiritual future was radically different to the colonial, westernised, world. We can position Aurobindo then as a thinker who was decidedly epistemically different – he certainly recognised his worldview was different to the 'Western code' which he resisted throughout his life, whether through political nationalism or spiritual universalism. Thus Aurobindo's 'spiritual' worldview was always *political*. It was not always progressive or egalitarian, as his writing in *The Indian Renaissance* shows, but it was never purely about retreating to an 'inner' world of disengagement in order to resolve spiritual dilemmas.

Conclusions

The anticolonial spaces which Aurobindo shaped were exceptionally dynamic. From setting out a manifesto for the idea of passive resistance and organising secret societies at the heart of Bengal's violent anticolonialism, to creating an alternative worldview based on a new version of yoga, the spaces which Aurobindo shaped show how anticolonialism is much more than simply *ressentiment*. This chapter has argued that, alongside the nationalist spaces of *Swadeshi* secret societies in Bengal, Aurobindo's retreat into 'the spiritual' was not simply a retreat from the outside world into an 'inner' one, and instead show how we must engage with a more nuanced understanding of how the 'minor' politics of individuals like Aurobindo force a recognition of the 'Western codes' which continue to shape our values and attitudes. Aurobindo then was undoubtedly thinking *politically*, even if he had withdrawn from the *Political*.

This is not to say that Aurobindo's universalist utopianism is an exemplar of how to conduct or imagine alternatives to the colonial system. As we saw above with Namakkal's work on Auroville, the utopian space of one group can often lead to the production of unequal and neocolonial relations for another, and Aurobindo's worldview is decidedly unipolar compared to the more heterogeneous framings which would be more palatable today. However, thinking about Aurobindo 'decolonially' does provide a way to think about the limits of colonial knowledge as they were deployed in India in the early twentieth century, and especially highlights the limits of what the colonial authorities would or could see as dangerous or anticolonial forms of behaviour. It is clear that his ideas radically exceeded the bounds of what was deemed to be politically possible according to the norms of his time, and thus his ideas were decolonial as they pushed back against the sensibilities of coloniality/modernity. Thus, they expand our notions of how anticolonialism can be productive of alternative codes and frames which in turn are generative of differential spatial relationships. As Aurobindo's life shows, individuals can often embody anticolonialism in very different ways within the course of their life. In the next chapter, I shift to a final member of the Pondicherry 'Gang', who had left the city before Aurobindo even arrived, M.P.T. Acharya, whose life course was equally diverse.

References

Alfassa, M. (1968) *The Auroville Charter* https://www.auroville.org/contents/1 (accessed 26 November 2018).

Ambirajan, S. (1995a). Human values and consciousness: towards a new social order in the light of Sri Aurobindo, Part 1. *Journal of Human Values* 1 (1): 127–138.

Ambirajan, S. (1995b). Human values and consciousness: towards a new social order in the light of Sri Aurobindo, Part 2. *Journal of Human Values* 1 (2): 249–264.

Anderson, B. (2006). Becoming and being hopeful: towards a theory of affect. *Environment and Planning D: Society and Space* 24 (5): 733–752. https://doi.org/10.1068/d393t.

Bose, S. (2007). The spirit and form of an ethical polity: a meditation on Aurobindo's thought. *Modern Intellectual History* 4 (1): 129. https://doi.org/10.1017/S1479244306001089.

Candy, C. (1994). Relating feminisms, nationalisms and imperialisms: Ireland, India and Margaret Cousins's sexual politics. *Women's History Review* 3 (4): 581–594. https://doi.org/10.1080/09612029400200066.

Chakrabarty, D. (2000). *Provincialising Europe: Postcolonial Thought and Historical Difference*, 2e. Princeton, NJ: Princeton University Press.

Chatterjee, P. (1993). *The Nation and its Fragments*. Princeton: Princeton University Press.

Dirks, N.B. (2001). *Castes of Mind: Colonialism and the Making of Modern India*. Princeton: Princeton University Press.

Dwyer, C. (2016). Why does religion matter for cultural geographers? *Social & Cultural Geography* 17 (6): 758–762. https://doi.org/10.1080/14649365.2016.1163728.

Featherstone, D.J. (2009). Counter-insurgency, subalternity and spatial relations: interrogating court-martial narratives of the Nore mutiny of 1797. *South African Historical Journal* 61 (4): 766–787.

Featherstone, D.J. (2012). *Solidarity: Hidden Histories and Geographies of Internationalism*. London: Zed Books.

Featherstone, D. (2013). Black internationalism, subaltern cosmopolitanism, and the spatial politics of antifascism. *Annals of the Association of American Geographers* 103 (6): 1406–1420. https://doi.org/10.1080/00045608.2013.779551.

Featherstone, D. (2019). Reading subaltern studies politically: histories from below, spatial relations, and subalternity. In: *Subaltern Geographies* (eds. T. Jazeel and S. Legg), 94–118. Athens, GA: University of Georgia Press.

Gandhi, L. (2006). *Affective Communities: Anticolonial Thought, Fin-de-Siecle Radicalism, and the Politics of Friendship*. London: Duke University Press.

Ghose, A. (1995). *Sri Aurobindo and the Freedom of India* (eds. C. Poddar, M. Sarkar and B. Zwicker). Pondicherry: Sri Aurobindo Ashram.

Ghose, A. (1997). *The Renaissance in India*. Pondicherry: Sri Aurobindo Ashram.

Ghose, A. (2002). *Bande Mataram: Political Writings and Speeches 1890–1908. Complete Works of Sri Aurobindo*, vol. 6 and 7. Pondicherry: Sri Aurobindo Ashram.

Gidwani, V. (2006). What's left? Subaltern cosmopolitanism as politics. *Antipode* 38 (1): 8–21.

Haggis, J., Midgley, C., Allen, M., and Paisley, F. (eds.) (2017). Interfaith, Cross-Cultural and Transnational Networks. In: *Cosmopolitan Lives of the Cusp of Empire*, 1860–1950. Basingstoke: Palgrave Macmillan.

Hardiman, D. (2013). Towards a history of non-violent resistance. *Economic and Political Weekly* XLVIII (23): 41–48.

Heehs, P. (2006). The uses of Sri Aurobindo: mascot, whipping-boy, or what? *Postcolonial Studies* 9 (2): 151–164. https://doi.org/10.1080/13688790600657827.

Heehs, P. (2008). *The Lives of Sri Aurobindo*. New York, NY: Columbia University Press.

Hill, L.J. (2015). More-than-representational geographies of the past and the affectivity of sound: revisiting the Lynmouth flood event of 1952. *Social & Cultural Geography* 16 (7): 821–843. https://doi.org/10.1080/14649365.2015.1026927.

Holloway, J. (2011). Tracing the emergent in geographies of religion and belief. In: *Emerging Geographies of Belief* (eds. C. Brace et al.), 30–52. Newcastle-upon-Tyne: Cambridge Scholars Publishing.

Holton, R.J. (2002). Cosmopolitanism or cosmopolitanisms? The universal races congress of 1911. *Global Networks* 2 (2): 153–170. https://doi.org/10.1111/1471-0374.00033.

Jazeel, T. (2015). *Matrimandir, Auroville*. Society and Space. Available at: http://societyandspace.org/2015/08/19/matrimandir-auroville-tariq-jazeel.

Kaiwar, V. (2015). *The Postcolonial Orient: The Politics of Difference and the Project of Provincialising Europe*. Chicago: Haymarket.

Lorimer, H. (2005). Cultural geography: the busyness of being 'more-than-representational'. *Progress in Human Geography* 29 (1): 83–94. https://doi.org/10.1191/0309132505ph531pr.

Mahajan, G. (2013). *India: Political Ideas and the Making of a Democratic Discourse*. London: Zed Books.

Mandair, A.-P. (2009). *Religion and the Specter of the West: Sikhism, India, Postcoloniality and the Politics of Translation*. New York: Columbia University Press.

Manjapra, K. (2010). Introduction. In: *Cosmopolitan Thought Zones: South Asia and the Global Circulation of Ideas* (eds. S. Bose and K. Manjapra), 1–19. Basingstoke: Palgrave Macmillan.

Menon, D.M. (2015). Writing history in colonial times: polemic and the recovery of self in late nineteenth-century South India. *History and Theory* 54 (4): 64–83. https://doi.org/10.1111/hith.10779.

Mignolo, W.D. (2011). *The Darker Side of Western Modernity: Global Futures, Decolonial Options*. London: Duke University Press.

Mignolo, W.D. (2018). Decoloniality and phenomenology: the geopolitics of knowing and epistemic/ontological colonial differences. *Journal of Speculative Philosophy* 32 (3): 360–387.

Mohanty, S. (2015). *Cosmopolitan Modernity in Early 20th Century India*. Oxford: Routledge.

Mukherjee, S. (2018). *Indian Suffragettes: Female Identities and Transnational Networks*. Oxford: Oxford University Press.

Namakkal, J. (2012). European dreams, Tamil land: Auroville and the paradox of a post-colonial utopia. *Journal for the Study of Radicalism* 6 (1): 59–88.

Prince, B. (2017). *The Integral Philosophy of Aurobindo*. Oxford: Routledge.

Ramaswamy, S. (2010). *The Goddess and the Nation: Mapping Mother India*. Durham, NC: Duke University Press.

Ravi Vaithees, V. (2015). *Religion, Caste and Nation in South India: Maraimalai Adigal, the Neo-Saivite Movement, and Tamil Nationalism 1876–1950*. New Delhi: Oxford University Press.

Raza, A., Roy, F., and Zacharia, B. (2015). Introduction. In: (eds. A. Raza, F. Roy and B. Zachariah), xi–xli. New Delhi: SAGE Publications India.

Rowlatt Committee (1918) 'Sedition Committee Report'. Edited by G. of I. Home Department. Calcutta: Government of India.

Sarkar, S. (2010). *The Swadeshi Movement in Bengal 1903–1908*, 2e. Ranikhet: Permanent Black.

Shilliam, R. (2015). *The Black Pacific: Anti-Colonial Struggles and Oceanic Connections*. London: Bloomsbury.

Wagner, K. (2019). *Amritsar 1919: An Empire of Fear and the Making of a Massacre*. London: Yale University Press.

Wolfers, A. (2016). Born like Krishna in the prison-house: revolutionary asceticism in the political ashram of Aurobindo Ghose. *South Asia: Journal of South Asian Studies* 39 (3): 525–545. https://doi.org/10.1080/00856401.2016.1199253.

Zachariah, B. (2015). Internationalisms in the interwar years: the travelling of ideas. In: *The Internationalist Moment* (eds. A. Raza, F. Roy and B. Zachariah), 1–21. New Delhi: Sage Publications.

Chapter Seven
The 'International' and Anarchist Life of M.P.T. Acharya

Introduction

On 8 November 1911, Mandayam Prativadi Bhayankaram Tirumala (M.P.T.) Acharya wrote to V.V.S. Aiyer in Pondicherry. Acharya had recently travelled to Constantinople from Germany on the advice of fellow revolutionaries Ajit Singh and 'Chatto' – Virendranath Chattopadhyaya – and was anxious to see if Aiyer thought he had made the correct decision: 'What do you think of my present location? Is it altogether disagreeable to you? Do you think it is but a perpetuation of my life in Germany, a waste, and no help to our future and propaganda? Please write me frankly' (GO 1014 Judicial, TNA). Despite his strategic uncertainty, Acharya did seem to know roughly what was going in wider revolutionary circles across the world – rumours of Aiyer's involvement in the murder of Collector Ashe and whether he would be arrested had reached Europe, and Acharya wonders in the letter about whether Aiyer is still free. Somewhat astonishingly, Acharya then goes on to discuss an international attempt to foment conspiracy and revolution with surprising openness. Given the Ottoman Empire's war with Italy which was taking place in North Africa at the time, Acharya was keen to see what was happening in India:

> I read that Indian Mussalmans [*sic*] are collecting money for Turkey. Is that true? I can't believe – for has not England prohibited "either party" with men, money or ammunition. How dare they disobey and incur the displeasure of "Our Emperor". I read also that the Mussalman papers of the Punjab have written somewhat strongly

Geographies of Anticolonialism: Political Networks Across and Beyond South India, c. 1900–1930, First Edition. Andrew Davies.
© 2020 Royal Geographical Society (with the Institute of British Geographers). Published 2020 by John Wiley & Sons Ltd.

"exciting race and religious hatred" since the war began and the Lieutenant-Governor has warned them. Between the lines, I read the race and religious hatred is against [Christians] and [Christianity] – Until now, I believe I am right. If so, that is good. Are the Hindus doing anything to help Turkey, say, by way of finance[?] – the idea should be set in motion by our revolutionaries that the Hindus should join the Muhammadans. In fact, we ought to have taken the lead before the Muhammadans, to give the Turko-Italian War an anti-European turn as far as possible, and to bring such Moslems [sic] as are patriotic enough to the Hindu Revolutionary Party. This has been a good opportunity and is still good.

Acharya's words here give a fascinating insight into the transnational world of revolutionary plotting that existed prior to World War One. Networks of knowledge and information were international, allowing news from India to reach revolutionaries like Acharya in Europe, but were fragmented. However, it is also clear that the Hindu-centredness to the spiritual revivalism of the *Swadeshi* movement was hitting its limits with people like Acharya, who were wondering how to build solidarities across communal lines. This was something that people like Aurobindo were, if not disinterested in, then certainly did not see as a priority (Heehs 2008), although Bharati's more liberal position was decidedly Pan-Indian. However, Acharya would change his position radically over the coming years, and by 1920 had declared himself an anarchist, and he travelled extensively throughout revolutionary and left-wing circles in Europe and beyond throughout his life, before returning to India and living in Bombay until his death in 1954.

Acharya's letter was intercepted by the British and is now contained in a partially related file about whether British Police should be retained in Pondicherry – the evidence that someone like Acharya was contacting Aiyer in Pondicherry was a symbol of the city's role in organising revolutionary anticolonial activities, although, as we have seen, by November 1911 this importance was in decline. The interception of the letter shows that by 1911, British (and other European) intelligence agencies had begun to establish effective networks to monitor revolutionary networks at home and abroad (Home, Political, Branch B, December 1909, No. 37, NAI). Networks of violent revolutionaries had been a particular concern across Europe, especially given the various acts of Anarchist 'Propaganda of the Deed' such as the assassination of King Carlos of Portugal and his son Prince Luis Filipe in Lisbon in 1908. As Anderson (2007) has pointed out, the transnational effects of European anarchism spread globally and inspired anticolonial movements across the world. Indian revolutionary movements were no different here – Hem Chandra Das, the key organiser alongside Barin Ghose of the Maniktala conspiracy which caused the trial of Aurobindo discussed in the last chapter, travelled to Europe in 1907 to be trained in bomb making. In Paris, he was introduced (probably by Emma Goldman) to 'Libertad' – the anarchist Joseph Albert – who likely provided some training in bomb-making (Heehs 1992, 2008).

THE 'INTERNATIONAL' AND ANARCHIST LIFE OF M.P.T. ACHARYA   139

There has been increased attention on these revolutionary movements in recent years, but M.P.T. Acharya is a little known but fascinating figure in the international Indian revolutionary movement that emerged in the first half of the twentieth century. Until recently, the only substantive text was a partial and incomplete autobiography which included a short but often error-filled biography (Acharya 1991), or the hard to find account of his 'life and times' written by CS Subramanyan (1995) which focusses on the earlier Bolshevik period of Acharya's life. Apart from that, he appears fleetingly in the biographies of notable revolutionaries like Virendranath Chattopadhyaya (Barooah 2004), and his name crops up in numerous secret service files, but often with little detail. However, Acharya was known to be a significant revolutionary who had been under British surveillance for some time. What makes him of further interest is the evolution in his political thought during the course of his lifetime. This has meant that in recent years, Acharya has become more visible recently – Maia Ramnath devotes a section of *Decolonising Anarchism* (2011a) to him which affords him equal importance as better known revolutionaries like Bhagat Singh, and since then, Bernstein (2017) has written about Acharya's time in Soviet Russia, whilst Ole Birk Laursen has conducted extensive work on trying to trace Acharya's writing, some of which is in his edited collection *We Are Anarchists* (Acharya 2019) and represents an important contribution to the study of anarchist internationalism's intersections with Indian anticolonialism.

This chapter uses the life of Acharya to stitch together some of the transnational connections which bound together various revolutionary movements across the globe in the first half of the twentieth century. As we shall see, Acharya's biography is complex and could be the subject of a number of books in itself. Instead of writing a narrative about this, I use work in the history of these revolutionary movements alongside geographical studies of social movements to understand some of the 'international' spaces which were produced by revolutionaries like Acharya. As Acharya's letter above shows, these spaces further illustrate the connections between Pondicherry and other revolutionary spaces across the world during the first two decades of the twentieth century.

However, they also are indicative of the variegated revolutionary networks which evolved over time and which people like Acharya had to negotiate, often at considerable cost to themselves and those around them. For example, whilst Acharya's life story ended in hardship and poverty in Bombay, his close friend and associate Chatto, who remained in Soviet Russia after Acharya had renounced Bolshevik Communism, was killed in the Stalinist purges of the 1930s. Whilst the revolutionaries faced these very real consequences for their political choices, the spaces of the international revolutionary movement were also cosmopolitan spaces, where interracial marriage was accepted at a time when it was not commonplace. This also extends to Acharya, whose marriage to the Russian artist Magda Nachman opens up these networks to a discussion of the 'politics of friendship' which both defined the process of building anticolonial solidarity, but also defined the intersections between the spiritual and political in the last chapter. Similarly to Aurobindo in the last chapter, this tells us something about how these political and personal relationships were gendered, as the opportunities for these kinds of relationships were much more likely to be open to Indian men than women. However, it is clear that Indian anticolonialists in Europe occupied a socially and politically cosmopolitan world which, while dangerous, opened up a world of radical new possibilities.

To explore some of these themes, I draw upon the vibrant work done in the geography of social movements. This scholarship is now wide ranging, but for the purposes of this chapter, I particularly focus on issues related to the study of transnational or international spaces within these movements. From here, I move on to discuss the various international 'networks' of Indian revolutionaries which were present in the first 30 years of the twentieth century. Following this, I discuss how Acharya's life intersected with these and argue that looking at Acharya provides an interesting counterpoint to some of the dominant narratives of Indian anticolonial organising in this period.

Geographies of Social Movements and Internationalism

The study of social movements has proved a vibrant area of study for geographers. Whilst much of the content of this book so far has focussed on contentious forms of politics in the form of anticolonial resistance, in this chapter it is worth utilising the social movements approach, and geographers' engagements with it, in order to understand some of the transnational linkages which were at work in the networks which Acharya traversed.

The study of social movements as a distinct form of political activity emerged post-1968 as sociologists became interested in the ways in which non-formal or civil society actors attempted to shift dominant debates in the political mainstream. The resultant corpus of literature, framed as either 'Social Movement Studies' or 'Contentious Politics' is indebted to work by the likes of Charles Tilly (Tilly and Wood 2012), Sidney Tarrow (Tarrow 1994) and Doug McAdam (McAdam, Tarrow and Tilly 2001) amongst many others in establishing a framework by which 'social movement actors' seek to establish some form of collective group and/or identity in which to resist a powerful organisation – traditionally, but not exclusively, the state.

Given this framing, geographers have interrogated a number of core areas of concern. For instance, some of these could include how do small-scale or 'local' social movements try to levy or 'upscale' their protests? How has globalisation allowed movements to build spatially extensive networks? Relatedly do movements build solidarities across the differences that they encounter within these practices? In their edited collection on the spatiality of social movements, Nicholls, Miller and Beaumont (2013) identify five trends that occur within geographical studies of social movements. Firstly, various papers have examined the ways in which social movements are placed – either routinely enacted through place, from the role of place in creating political subjectivities, the importance of proximity to building long-term relationships between social movement actors, but also can act as a hindrance to movements which become too 'localised' – what David Harvey (drawing on Raymond Williams) termed Militant Particularism (Harvey 1995). Secondly, the territorial construction of space matters for social movements, whether in resisting territorial impositions and boundary making practices, such as the No Borders movement, or through the ability for social movements to claim territories and rework them to their needs, such as the Occupy! Movement (Halvorsen 2015). Thirdly, issues about how power and contestation are scaled have been of importance. Recognising that scale is relationally produced, social movement activity often enfolds a variety of overlapping spatial scales in order to make their cases, from municipality to international agencies (Miller 2000). How we understand the uneven and relationally produced scaling of power across these different scales is a crucial aspect of understanding how social movements make sense of their geographies.

Fourthly, in the advent of the technological and financial globalisation of the 1980s and 1990s, and the emergence of networked forms of contestation such as the *Ejército Zapatista de Liberación Nacional* or 'Zapatista' rebellion in Mexico in the 1990s, the ability of social movements to construct spatially extensive networks or information, people and other materials became of increased importance. This has overlaps with some of the discussions of scale, especially where building transnational networks was seen as an effective way to 'jump' scales, so 'local' movements could act at global levels. However, networked approaches to space more generally, especially given the turn towards Actor-Network Theory in geography in the 2000s, often emphasised the connections and flows between and across space which tended to downplay or 'flatten' some of the power inequalities present in the world, and so geographers attempted to show how networks were relationally constituted both within and across space – for example, in Nicholls's study of the social movement networks which exist in urban spaces (Nicholls 2008). Lastly, geographers have attempted to respond to critiques of the networked approach, through more attention to how 'entangled' power relations occur (Sharp et al. 2000). This is in line with shifts towards relational geographies, but is also representative of how geographers have begun to challenge the use of singular spatial frames to understand complex phenomena and instead think about how different spatialities co-constitute each other (Jessop, Brenner and Jones 2008; Jones 2016).

The debates over how best to think through these complex spatialities have been widespread (see Nicholls 2007; Leitner, Sheppard and Sziarto 2008; Koopman 2015 to name only a few), but geographical approaches to social movements have carved out a distinctive approach to these movements (Routledge 2017). They are also important as they have provided a way for geographers working at the intersections of history and politics, as, despite the emergence of the discipline of social movement studies largely happened post-1968, understanding their historicisation has always been important (Tilly and Wood 2012). In particular for the purposes of this chapter, understanding how contentious political activities were spatially extensive can challenge the idea that it is only since the 1980s and 1990s that social movement networks have been important. For example, David Featherstone's work has been important in challenging David Harvey's vision of militant-particularism as a local issue, instead calling for more relational and spatially 'stretched' accounts of subaltern struggle across space (Featherstone 2005). These spatially stretched examples of solidarity 'from below' have formed a cornerstone of Featherstone's work, which has been committed to thinking through the entanglements of these spatial connections, which, as previously noted, has been a core part of geographers' response to the 'flattened' world of networked spatial framings (Featherstone 2012, 2017). Much of my own previous work on the varieties of anticolonial social movement-style struggles of the past and the present (Davies 2012, 2013) has tried to work through the entangled spaces of political activity to show how various spatialities interlock and are contested within social movement and anticolonial space. Indeed, this book and its empirical content on Southern India is an attempt to show the complex relational geographies produced through anticolonial forms of politics, and which incorporate many of the spatial frames and imaginaries which Nicholls, Miller and Beaumont attribute to social movements.

The spatial framing and vocabulary of these debates is still important. There has been a tendency where the term 'transnational' has been supplanted to an extent by a focus on the 'translocal', which has done useful work in shifting debates away from the national scale to the urban (McFarlane 2009, 2011) or other situated locales. However, within historical geography, especially around the interwar era and the first half of the twentieth century more broadly, the international has become an important focus. The international was both an important term which came to prominence in the interwar period, with the growth of organisations like the League of Nations or the Third Communist International, but has been used by geographers to explore some of the 'entangled' geographies which were at work during this period. As Hodder, Legg and Heffernan (2015) argue in a recent introduction to a special issue on historical geographies of internationalism, the international's relationship to other scalar categories – from the imperial, colonial, the national, the local – is contextually variable and often unclear. The term is suggestive of a relatively elite process of interaction between national level actors – especially if thought through in terms of international organisations, conferences and meetings (such as the UN or League of Nations, for example). However, the international, and more widely, the political project of internationalism, has come under scrutiny and been challenged in this regard. This chimes with the research discussed in Chapter Three which explored how internationalism in the interwar period was a space by which various actors were involved in ongoing processes by which their political identities and subjectivities were formed, and which made the application of categories such as 'fascist' or 'socialist' in their twenty-first century understandings problematic (Raza, Roy and Zacharia 2015; Zacharia 2015).

In his work on the non-state based spaces of 1949 World Pacifist Conference, Jake Hodder has shown how these spatially extensively entanglements occur within 'internationalist' political spaces. As Hodder shows, notionally 'internationalist' spaces like the Conference become grounded in the locale in which they are enacted – in this case, India 'the land of Gandhi', but also in Santiniketan, the town in Bengal now indelibly associated with Rabindranath Tagore, the Bengali poet and hugely influential cultural figurehead (Collins 2012).[1] As Hodder notes, the international was not 'a given category or scale, but a way of encasing the different conceptions of the world which were tied to the places in which it was debated and sustained' (Hodder 2015). The international then always exists in a tension where it ceases to be truly international once it 'lands' or is 'placed' in the very spaces by which it is coterminously being produced, debated and established. As Hodder shows, internationalist political geographies then share many similarities with the entangled and relational understandings of how social movements are constructed. The international is important space to understand how radicals like Acharya made sense of their struggles. However, as the previous discussion shows, 'international' space is always a project that is unfinished and is articulated differently according to different actors. As Stephen Legg (2014) has argued, the challenge is often to understand how historical actors saw the international and sought to make use of it to their advantage – the international was differentially articulated and experienced by those involved in the League of Nations compared to those in the Communist International, or even in Aurobindo and Mirra Alfassa's spiritually universalist internationalism in Pondicherry/Auroville.

However, I have deliberately tried to avoid categorising Acharya rigidly as an 'internationalist' (hence the scare quotes in the title of this chapter). The work done recently by geographers of social movements and of internationalism means that it is appropriate to use these terms to understand Acharya's life – his letter to Aiyer at the start of this chapter shows him thinking 'internationally' in trying to mobilise Pan-Asian and inter-communal revolution based on his 'grounded' circumstances in Constantinople. However, given the changes in political inclination and attitude which he lived through – he identified firstly as a nationalist revolutionary and then later as anarchist rather than as an internationalist for most of his life – I instead use the international as a synonym to understand the ways in which numerous spatial and scalar processes were enrolled into his life and how he lived through these processes. This is important as it continues to

[1] To many Tamils, Tagore is seen as an equivalent to Subramania Bharati, despite Tagore's far greater fame (see Venkatachalapathy 2018).

show how diverse the strands of internationalism were in the interwar period. Given the often byzantine networks that Acharya was navigating, before turning to look specifically at his life, we need to set out what was happening in Indian revolutionary movements globally during his lifetime.

Geographies of Indian Inter/Transnational Revolution and the Inter War Period

Even though the presence of the Indian revolutionary anticolonialists abroad has been known for a long time, recently, the recognition of the importance of them to the story of Indian anticolonialism has fundamentally changed the story of much of the interwar years. This has occurred in two ways, firstly and most simply has been the increase in recognition attributed to revolutionary ideas in the Indian anticolonial movement more generally (Maclean and Elam 2013; Maclean 2015; Elam and Moffat 2016). Secondly, has been the increased acknowledgement, driven by the turn towards globalised and transnational histories, of international and cosmopolitan tendencies across the entire freedom struggle, of which revolutionary activities were only a part (Bose and Manjapra 2010; Bayly 2012; Mukherjee 2018). These two tendencies have significant overlaps, and so in the following discussion I do not attempt to create a false separation and instead explore them in a relatively chronological order. These ideas have been briefly mentioned in some of the previous chapters, but I expand on them in more depth here, both to link them to the ideas about social movements and internationalism discussed previously, but also to provide a better context to Acharya's life in the coming sections.

We have already seen how *Hind Swaraj*, the archetypal Indian anticolonial text, was written by Mohandas Gandhi as a response to the activities of the extremists in the Indian National Congress (INC). On the one hand, knowledge of Gandhi's activism in South Africa was widespread in India itself, and he was also aware of the debates about what was happening in India. The debates about passive resistance and *satyagraha* which Gandhi was developing were drawing on and critiquing Aurobindo's 'extremist' version of passive resistance (Ghose 2002), which was open to violence if necessary. However, Gandhi was also writing about the revolutionary networks of Indian extremists in Europe, and here more explanation of their activities is needed.

Whilst in India, the Partition of Bengal had mobilised *Swadeshi* activism, internationally, following the Japanese navy's overwhelming defeat of Imperial Russia at the Battle of Tsushima in 1905, there was increased belief that the age of European 'imperialism' was coming to an end, with the possibility of an ascendant Asia taking its place (see Hyslop 2011 for a link between these events and *Hind Swaraj*; also, Mishra 2013 discusses the importance of Tsushima for differing forms of Pan-Asianism). The ability of Indians and other subjects of Empire to travel relatively freely, and to establish spaces by which the colonies became visible in spaces across Europe. As Bayly (2004) argued, the first decade of the twentieth century was a period of marked intensification in global political and economic exchanges. Whilst the movements and groupings of Indians overseas were by no means new (Visram 2002; Bald 2013), they did begin to create new forms of diasporic space (Brah 1996) and were vociferous enough to be able to make claims about their conditions, whether at home or in the diasporic space of the metropole (Boehmer 2015). These diasporas were however still relatively elite spaces, with the majority of those Indians present in the United Kingdom and across Europe being students (Fischer-Tine 2007).

In 1905, India House was opened by Shyamji Krishna Varma as a home for Indian students in Cromwell Road, Highgate, London. In the 1800s, Krishna Varma had been a colonial official, but having become disillusioned with British rule, moved to London in 1896 and used his wealth to promote Indian Home Rule. As well as opening India House, alongside the British radical socialist Henry Mayers Hyndman, he founded an Indian Home Rule Society, and started publishing *The Indian Sociologist*. This paper was soon the mouthpiece of radical revolutionary politics, often drawing upon the anti-imperial ideas of Herbert Spencer (Kapila 2007) and was one of the papers which was being imported surreptitiously into Pondicherry within a few years. India House swiftly became a centre for radical agitation, especially as Krishna Varma was able to provide funds for scholarships for gifted students, including V.D. Savarkar, the 'father' of *Hindutva* who has been discussed previously (Fischer-Tine 2007; Tickell 2012). V.V.S. Aiyer was also actively involved in India House and in the publication of *The Indian Sociologist*. It was one of the students based at India House, Madan Lal Dhingra, who assassinated Sir William Curzon Wyllie at an event organised by the National Indian Association at the Imperial Institute on 1 July 1909, and as discussed in Chapter Two, it was this act that spurred Gandhi to write *Hind Swaraj* in the days following the assassination whilst travelling back to South Africa and having encountered the extremists at India House during his time in London.

The inevitable crackdown on India House after Curzon Wyllie's assassination meant that London soon stopped being the centre of revolutionary organising. Krishna Varma had already left in 1908 to relocate to Paris, who was followed by Bhikaji Cama (often called Madame Cama) who arrived from London in 1909 and established the Paris Indian Society. Cama is an important figure in Indian nationalist history who deserves renewed scholarly attention. A member of India House, in 1907, she unfurled the 'Flag of Indian Independence' at the International Socialist Conference in Stuttgart, co-designed by Cama and Krishna Varma, but was left disappointed by the unwillingness of the Conference to address matters related to 'subject peoples' in any depth. Having been in India House in London, her base in Paris provided an important space for revolutionary activity to continue, and soon the likes of V.V.S. Aiyer were based in Paris alongside her.[2] As well as continuing the publication of the *Indian Sociologist*, Cama organised the publication of the Geneva-based *Bande Mataram*. It was from her or her Parisian associates that shipments of these to Pondicherry, discussed in Chapter Five, originated. It was also Cama who suggested Paul Richard would be sympathetic to the goals of the Pondicherry 'Gang' if he won election as the Deputy to Pondicherry, and thereby encourage his first visit there in 1910 and setting off the chain of events that would lead to Aurobindo's encounter with Mirra Alfassa in 1914 (Yadav 1992).

Cama's role as a facilitator of what was an undoubtedly masculinist internationalist revolutionary space is something that is worthy of more geographical attention. Whilst Paris provided one space where revolutionary activity could continue under the leadership of Cama, there were two other trends which emerged in the movement. One was the emergence of the '*Ghadar*' movement, and the second was the confluence of various groups of revolutionaries with international Socialism and Communism.

The '*Ghadar*' movement is notable as it spread the message of anticolonialism and anti-imperialism to North America, but also acquired a global reach. There was a notable community of anticolonialists at work prior to *Ghadar*, and there existed a degree of Indo-Irish cooperation in New York, and in an extension of the existing India House networks, the *Indian Sociologist* found space to share stories with the Irish Nationalist newspaper the *Gaelic American* (Fischer-Tine 2007), and Bhikaji Cama was invited to give a lecture tour in 1907 (Yadav 1992). However, whilst *Ghadar* was closely associated with some members of the Paris network, such as through the likes of Muhammed Barakatullah who was involved in a huge variety of revolutionary organising, it swiftly expanded to become much more dynamic and spatially extensive. In its original form, *Ghadar* was mobilised around a group of revolutionaries on the West Coast of the USA, notably Lala Har Dayal, but also Taraknath Das, Ram Nath Puri and others, who agitated amongst the disaffected Indian migrant populations of North America and began to make a call for an armed insurrection against British rule – *Ghadar* was the title of the movement's publication and translates as 'Revolt'. At the first meeting of the Hindi Association formed in Portland in May 1913, Har Dayal summarised the goals of the emergent movement:

> Do not fight the Americans, but use the freedom given that is available in the US to fight the British; you will never be treated as equals by the Americans until you are free in your own land; the root cause of Indian poverty and degradation is British rule and it must be overthrown, not by petitions but by armed revolt. (cited in Chandra et al. 1989, p. 149)

---

[2] Aiyer eventually left and arrived in Pondicherry in December 1910 to organise the revolver practice that led to Robert Ashe's death.

The similarities in argumentation between this and the extremists in the INC are obvious and show the close connections between many of these organisers, but as Maia Ramnath has shown in her study of *Ghadar* (Ramnath 2011b), it was a multifaceted movement which incorporated various strains of politics, from syndicalism through to republicanism. Often seen as a Punjabi movement as many of the Indian communities on the West Coast of America had emigrated from there, the group was in fact multi-ethnic and contained leaders from a variety of different religious/communal groups. *Ghadar* also spread globally from the United States to South America, the Middle East, Europe, South-East Asia and Japan, and is one of the first examples of a global political network, especially one based on the anticolonial imperative.

Ghadar's attempts to foment a revolution in India, principally in the Punjab, filtered into a series of attempts by the German state to foment rebellion in the British Empire in World War One. This was coordinated by both the *Ghadar* members in the United States, but through revolutionaries in Berlin such as Chatto and Acharya who attempted to smuggle weapons and counterfeit money to India. This 'Berlin Committee' formed part of a wider Hindu-German Conspiracy which sought to use the logic of 'the enemy of my enemy...' to use German resources and Indian revolutionaries to cause chaos in British colonies during World War One. *Ghadar* messages were involved in fomenting the Singapore Mutiny of various units of the British Indian Armed Forces in 1915 (Abraham 2015), and the establishment of a revolutionary 'Provisional Government of India' in Kabul in December 1915, which included *Ghadar* members like Barakatullah.

One of the most extraordinary events was what became known as the *Komagata Maru* incident in the spring and summer of 1914. This involved a steamship, the *Komagata Maru*, which was carrying Punjabi immigrants to Vancouver, Canada being turned away forced to return to Calcutta. The organiser of the voyage, a Sikh entrepreneur named Gurdit Singh based in Singapore, had identified with the *Ghadar* movement, but also arranged the transit of the ship as a way to circumvent restrictive Canadian immigration laws that stopped entry for individuals who had not arrived in Canada from their home country in one continuous journey, and which were largely seen as racist. The voyage of the *Komagata Maru* was well publicised, and the Canadian Government claimed that *Ghadarite* revolutionaries were on board. The ship sailed from India but had to stop in numerous ports on the way. When, having finally sailed from Hong Kong, the ship reached Vancouver, it was not allowed to dock. After protests across Canada and the United States, a few passengers were allowed to enter Canada, but the ship was forced to leave Vancouver Harbour and was forcibly escorted out of the harbour by Canadian naval vessels. The ship was now seen by the British as being a haven for revolutionaries, and when it docked in back in Calcutta in September and the supposed ringleaders resisted arrest, police opened fire and killed 19 people. The *Komagata Maru* provoked outcry and was a visible example of the racist policies of the Empire, and Canada in particular. The episode's connections to *Ghadar* meant that it was able to be used to publicise the racism at the heart of empire, and to connect this to the wider struggle for independence, and the incident remains an important part of Canada's reckoning with its colonial past today, especially for South Asian diasporic communities (Roy and Sahoo 2016; Almy 2018).

Ghadar's attempts to foment rebellion within India were ultimately unsuccessful. The public nature of many of the *Ghadarite* pronouncements and their widespread circulation in print meant that the colonial authorities were aware of many of the individuals taking part and could arrest many of them as they tried to enter India. Those that did make it to the Punjab found that there was not enough mass support for an outright rebellion to take place. However, *Ghadar* is hugely important in the study of inter/transnational Indian anticolonialism, as it was able to spread the ideology of anticolonialism globally and not just to elite audiences (Ramnath 2011b). Abraham's study of Singapore Mutiny (2015) has argued that dismissing these seemingly insignificant moments in counter-imperial history excludes the more diverse forms of insurgent identities which were being constructed in such places/events. To him, long-distance travel allowed the soldiers who mutinied in Singapore to construct forms of subjectivity that were based instead on notions of equality and emancipation rather than an adherence or desire for an independent nation state. Again, this highlights the indeterminate nature of 'politics' during this era, but it also shows how the ability to travel and compare across situations was a core aspect of the building of resistant political subjectivities (Ahuja 2010; Davies 2014).

Ghadar's multifaceted geographies led into a variety of further revolutionary connections, although the move into 'interwar' forms of internationalist politics meant significant changes to what was possible. Post-World War One and the Russian Revolution of 1917, as well as the turmoil unleashed by the end of the War, the opportunities for the utopian forms of socialism that had shaped people like Mirra Alfassa's life were rapidly closed down. The ascendancy of Bolshevism and Leninism meant that, as Gandhi (2006) has discussed, the 'utopian' moment of connection across difference which these spaces opened was short-lived and dismissed as 'immature' in the context of an expanding international communism. However, whilst this closed down opportunities for some connections across difference, for many amongst the international revolutionaries, the emergence of Soviet Communism and the USSR meant that there was a viable alternative to colonial rule. Many of those who had been involved in the Berlin Committee found their way into various currents of international socialism, communism and anarchism.

There were numerous connections between anticolonial revolutionaries and communism at this time, and not only with South Asians (White 1976; Derrick 2008; Høgsbjerg 2011; Adi 2013; Featherstone 2017). One of the most notable attempts by the Comintern to manage anticolonial activity was in the formation of the League Against Imperialism (LAI). Started with an International Congress against Imperialism and Colonialism in Brussels in February 1927, the organisation which became known as the LAI, was coordinated by the German Communist Willi Münzenberg in Berlin and was based there until 1933. Fredrik Petersson (2014) has argued that the LAI, and particularly Münzenberg, have been underserved by history as his later expulsion from the Communist Party meant he was treated as persona non-grata in many histories of the interwar Comintern. This situation is changing, with Prashad (2008) linking the LAI, and in particular the 1927 International Congress, to the emergence of political Third Worldism after the Bandung Conference of 1955.

Here is not the place to go into a detailed history of the LAI or international communism (see Petersson 2014 for a brief summary), but it is important to note that the location of its International Secretariat in Berlin in the 1920s and 1930s meant that it was based in a city with huge numbers of émigrés, and thus was able to act as a hub space though which a number of anticolonialists moved. Chatto, Acharya's friend, was heavily involved in the LAI and was a its General Secretary (Barooah 2004), but numerous other anticolonial, and increasingly nationalist, movements were present in Berlin. However, the LAI was also always existing in a tenuous relationship with the Comintern. For instance, after the Sixth Comintern Congress in Moscow in 1928 called for a united front in the colonial question, which in practice meant the closing down of any links with social democrats, pacifists and other groups, this effectively meant the limiting of what was possible for organisations like the LAI and stifled the cosmopolitan potential that had existed in the fin-de-siècle (see also Worley 2004). However, despite this, the LAI was able to fashion an important network of individuals, largely due to the presence of so many of them in Berlin. Thus, again, we can see the 'international' League becoming inherently placed through its location in Berlin. However, with the rise of Nazism and an increasingly hostile environment for Communism in Germany more generally, the LAI was unable to exist for long – the final meeting of its International Secretariat was on the 30 January 1933, the day that Hitler was appointed Reich Chancellor.

The LAI highlights how much the room for manoeuvre for radical anticolonialism had become restricted by the late 1920s. The huge societal changes that occurred globally post-World War One are important here, as well as the hardening of various positions on both the left and right of politics. In order to understand parts of these networks more fully, and to draw this back towards the Pondicherry 'Gang', the chapter now turns towards M.P.T. Acharya.

M.P.T. Acharya's 'International' Lifecourse

Acharya was born in Madras in 1887. As pointed out in the earlier chapters of this book, he was involved in the publication of *India* and *Bala Bharata* with Subramania Bharati in Madras until they were forced to move to Pondicherry in

1907. Unlike Bharati, who remained in Pondicherry until 1918, Acharya soon left for Europe. As was noted previously, the revolutionary activity taking place in centres like Paris and London was well known in India, and the prospect of being 'trapped' in Pondicherry probably held little appeal to Acharya. Unlike Bharati, who could work and produce his poetry and songs alongside his political writing, Acharya saw himself mainly as an active revolutionary at this point. Acharya arrived in London and became a member of India House, and immediately set about trying to learn the practices of anticolonial revolution.

His first notable attempt at taking part in revolutionary activity was in 1909, where, along with Sukhsagar Dutt, a fellow member of India House, he attempted to go to Morocco and join the Rif rebellion against Spanish colonialism. Writing in 1958, Acharya's companion Dutt described how this trip was inspired by V.D. Savarkar, who at the time was the leader of the India House group (cited in Bose 2002, p. 39). The trip, it seems, would have had the dual purpose of building sol-idarity between different anticolonial movements, but would also have given Acharya and Dutt front-line (literally) experience of fighting in a revolutionary guerrilla war. In his reminiscences, which were serialised in the *Mahratta* in the late 1930s, before being reproduced in the 1990s (Acharya 1991, p. 98), Acharya claimed more simplistically:

> I decided to go anywhere where I could live cheaply and also do some nationalist work. I was also nauseated by the "freedom" in the freest of countries. I would have returned to India already if I was sure no persecution would take place. But after the actions in England, persecution was more certain than [e]ver in India.

However, there was a degree of expediency to his move. Following Dhingra's assassination of Wyllie, Acharya was involved in protests against Dhingra's sen-tencing. V.D. Savarkar spoke up in support of Dhingra at meeting which was intended for Indians to speak out deplore the assassination. When a barrister (given as a Mr. Palmer by Acharya) attacked Savarkar, Acharya hit Palmer with a stick, and an altercation broke out. Acharya was, then, living on borrowed time in London, so getting out before he was arrested was probably sensible. Having sailed first to Gibraltar, it swiftly became clear that it would be difficult for the two men to reach North Africa. Steamship travel opened up the world to the revolutionaries, but also meant that, if the authorities knew which ship someone was on, it was relatively easy to trace their movements. Acharya's movements from London to Gibraltar were noted straight away (Home, Political, Branch B, December 1909, No. 37, NAI). Dutt's recollection of the time in Gibraltar in his letter 50 years later is simply that he and Acharya were 'not able to proceed to the rebel territory' (cited in Bose 2002, p. 39). Acharya is harsher in his reminiscences, claiming that Dutt was 'unwilling to go to a country and for a purpose which would not give the comforts of London student life' (Acharya 1991, p. 100).

Whether or not Acharya actually made it to North Africa is unclear. Dutt certainly didn't, but Acharya claims to have gone in his reminiscences, whilst the Weekly Report of the Director of Criminal Intelligence on 4 October 1909 in India noted that he had made it to Tangier with thirty shillings on 26 August, but was writing to V.V.S. Aiyer within two days confirming that the Rif plan could not be conducted (Home Political, Branch B, November 1909, Nos. 32–41, NAI). What is clear is that Acharya was too late – by the time that any attempt to act was made, any insurgency or guerrilla warfare by the Rif had been defeated. Having exhausted what meagre funds he had, Acharya retreated to Paris, via a brief stay in Portugal, which he thought a 'colony of British capital' (Acharya 1991, p. 112). This allowed him to link up with the various revolutionaries active in the Paris Indian Society, many of whom he knew from London. However, the attempted move to help the Rif was undoubtedly a failure, and the lack of planning and also any existing connection with the Rif prior to the attempt show the limits to the knowledge of the India House group at the time.

Moving to Paris immersed Acharya in a flurry of activity, and it is hard to keep track of his movements alongside the other revolutionary anticolonialists based there, as well as the activities he was conducting. It was reported in April 1911 that he had been sent to Rotterdam to learn engraving and printing, but then moved to Berlin 'for the purpose of spreading their propaganda amongst the few Indian youths there' (Home, Political (Deposit), April 1911, No. 7, NAI). Whilst the Department of Criminal Intelligence (DCI) put together reports or 'History Sheets' on notable individuals in Paris, Acharya is notable by his absence in having a report dedicated to himself, which indicates how the colonial authorities often thought of him as a marginal figure. The DCI's reports for the activities of the group cannot be treated as a wholly accurate record and are often incorrect – the file Home Political (Confidential) Branch A, August 1919, Nos. 44–52, for example believed that Acharya had moved from Pondicherry to Paris, rather than London – as well as the clear drawbacks of only being written from the perspective of the coloniser. However, a sense of the pace of life and the international networks which were present in Paris becomes clear:

On the 16 September 1910 orders were issued by the French Government prohibiting the holding of the Egyptian Conference[3] anywhere in France and it was decided to have it in Brussels instead. In the evening of 21 September the Indian and Egyptian extremists held a soiree at Palais Hotel, Champs Elysees, which was attended by Shyamji Krishnavarma, SR Rana and many other leading extremists. Next morning, most of those who attended went to the Egyptian Conference in Brussels. Here Miss Perin Naoroji[4] distinguished herself by singing some national

---

[3] An unknown conference of Egyptian nationalists.
[4] Perin Naoroji was the granddaughter of the moderate nationalist Dadabhai Naoroji (of 'Drain theory' fame).

songs. Madame Cama also was present and spoke on 23 September on female education. In the course of her speech she remarked on the way in which Indian prisoners were treated in India and how they were put into cages and places in the burning sun. She advocated a general rising of all Indians against the English, and the throwing of bombs at English officials in India. She was ruled out of order by Mr Keir Hardie, MP, who was presiding at that sitting of the Conference, when she opened the subject of the treatment women received in India. (Home Political, Branch N, August 1913, No. 61, NAI)

Although it is not clear exactly what this conference was, the mixing of Indian and Egyptian Nationalists alongside British Labour MPs is indicative of the cosmopolitan nature of these spaces. Acharya moved in these circles, but was moving rapidly from city to city at this time.

Having moved to Berlin, he left there for Constantinople in 1911, where he wrote the letter to Aiyer that was discussed in the introduction to this chapter. He did not remain there long as his hopes for revolutionary organising as expressed to Iyer in Pondicherry were not met. He then moved to New York and seems to have spent a few years there, before editing the Tamil edition of *Ghadar* in San Francisco for a short time in 1914 (as noted in Ole Laursen's introduction in Acharya 2019). In 1914, he returned to Berlin and worked with Chatto on the Indian Independence Committee, one part of the Hindu-German Conspiracy. In April 1915, he was part of a mission to the Middle East to try to secure the support of Muslims against the British, before returning to Constantinople. As Germany's interest in supporting extremist activities faded as the war dragged on, Chatto and Acharya travelled to Stockholm in 1917, where another cluster of anticolonial and socialist intellectuals had gathered, where he remained for most of the next few years. In 1919, Acharya met Lenin in Moscow, had attempted to set up the Indian Revolutionary Association in Kabul, but was soon expelled by the pro-British Emir, and then relocated to Tashkent. Here, he and his comrades established the Provisional All India Central Revolutionary Committee in August 1920, which was further enhanced by the formation of the Communist Party of India (CPI) there in October, with Acharya as Chairman.

However, by this time, Acharya's views were causing friction (Bernstein 2017). He disagreed with the Secretary of the CPI, M.N. Roy (something which was not unusual given Roy's often controlling behaviour), and was expelled from it in January 1921 for his support for non-communist positions (Laursen 2019). Returning to Moscow, Acharya attended Pyotr Kropotkin's funeral and met his future wife, the artist Magda Nachman before realising his non-conformism was increasingly unwelcome in Russia, and he and Nachman moved to Berlin in 1922. Once in Berlin, he swiftly renounced Bolshevism, declaring in a letter to Chittaranjan Das, a Bengali radical, that he was an anarchist (Laursen, n.d.). Later, Acharya's vehement rejection of Soviet Bolshevism becomes clear in an introductory article on 'What is Anarchism' that he wrote for Iqbal Singh and Raja Rao's *Whither India* in 1948:

> The capitalists at least have a common platform with the Bolsheviks on the state issue – and therefore both capitalists and Bolsheviks are the deadly enemies of anarchists. *The capitalists are individual or group Bolsheviks while the Marxians are collective capitalists.* The anarchists are against both forms of capitalism. Only the capitalists and Bolsheviks agree that Bolshevism is socialism, which the anarchists deny. They call Bolshevism the worst form of capitalism. (Acharya 2019 [1948], emphasis in original)

Acharya would have been aware of anarchist ideas for much of his time in Europe – the British anarchist Guy Aldred was closely affiliated with India House and was sentenced to prison in 1909 for arranging the printing of the banned *Indian Sociologist* (Laursen 2018),[5] so it is unclear exactly when he decided that anarchism was his core ideology. However, this shows how open spaces like India House were in the earlier phases of this international revolutionary 'movement'. We could therefore read India House and the individuals who were active in its social circuits as another space of cosmopolitan encounter which was discussed in the last chapter. Despite his suspicion of Bolshevism, Acharya remained involved in left movements throughout Europe, and when Chatto was involved in setting up the LAI's Congress in 1927, Acharya assisted him, but found the LAI's connection to the Comintern too problematic to continue engaging with the organisation long term (Laursen 2019).

Having settled in Berlin, Acharya worked doing translations and pieces of writing, as well as working though his ideas about anarchism. As was previously noted, around this time, Berlin was also a centre for radical organising, not only through the auspices of the LAI (Petersson 2014). We can then add Berlin to the list of spaces through which internationalist politics was 'grounded' (after Hodder 2015) and which shaped the world of people like Acharya. The fact that Acharya was writing in a variety of languages and for numerous sources meant that it was hard for the authorities to trace what he was doing. This has also contributed to his absence from the historic records, and it is only in recent years and in forthcoming publications that his ideas are becoming more widely known. The advent of Nazism meant that by the mid-1930s, Berlin was unsafe for the mixed race marriage of the Acharya-Nachman's, and after a number of failed attempts to get a passport to return to India due to his political activities, he was finally granted one and moved to Bombay in 1935, with Nachman following in 1936. In India, Acharya published his reminiscences in the *Mahratta*, and wrote about anarchism

---

[5] On a related note, see Paul Griffin's work on Clydeside which explores how the likes of Aldred played important roles in forging diverse working-class political alliances (Griffin 2018).

in an unsuccessful attempt to encourage the spread of its ideas in India – it is only recently that the intersections between anarchism and India have begun to be fully explored, and this also contributes to the long overdue recognition of Acharya (Ramnath 2011a). Magda Nachman did enjoy some success in the Bombay art scene (Bernstein, 2018, and see also the collection of Nachman Acharya's work collected at https://magdanachmanacharya.org/). However, publishing and writing was not a stable income for the couple, and after World War Two, Acharya was troubled with ill health, including tuberculosis as a result of sustained poverty. Magda Nachman Acharya died in Bombay in 1951, whilst Acharya died in 1953.

This synopsis of Acharya's life indicates that he was of one of the most dynamic and nonconformist characters to emerge in the Indian independence struggle. Whilst we can see threads of libertarian socialism and related political forms in many of the individuals in this book, Acharya embraced these ideas and was happy to declare himself wholeheartedly an anarchist. That Acharya had experienced life in Soviet Russia and had met Lenin means that he was aware, at least in part, of the limits to Soviet versions of anticolonialism, something which the likes of Jawaharlal Nehru took longer to recognise given his more formal and elite engagement with anticolonial internationalism through organisations like the LAI. After his death, his editor in the *Harijan* described him as a 'total believer in the doctrine of philosophical anarchism' which he had come to hold 'as a result of his long and arduous campaigning in foreign lands for the cause of India's freedom' (cited in Ramnath 2011a, p. 133). This again indicates the diversity of anticolonialisms which were mobilised through a connection to Pondicherry. Acharya's life overlaps with many of the internationalist geographies of the interwar period which are already relatively well known. However, exploring Acharya's life uncovers a further strand to these internationalism and pushes at the intersections between them.

Acharya flitted through the more clearly established and well-known spaces of anticolonialism, whether in the form of the CPI, or in his connections to the organisers of the LAI. However, his mutability in political orientation is important as this shows more clearly than most how he was negotiating this position. This again pushes at the minor political spaces which bridge across the formal *P*olitical categorisations which are often applied to anticolonialists as post hoc rationalisations (Raza, Roy and Zacharia 2015). The revolutionary networks which Acharya were involved in were as diverse as something like the alter-globalisation movement of the 1990s and 2000s. Whilst the majority of the individuals involved would, at least initially, have characterised themselves as 'nationalists' working for Indian independence, there were different counter-currents at work within these movements. From socialism and communism to Acharya's anarchism, we can see how anticolonialism was diverse and produced relationally in the spaces and places in which people like Acharya 'touched down'. Places like Pondicherry, Paris, Berlin, Gibraltar and Constantinople all provided different ways of envisioning how best to fight for and mobilise others to become involved in anticolonial work. Seeing

Acharya's life as relationally interconnected with other anticolonialists, as well as other individuals like his wife Magda Nachman, continues to emphasise the spaces of anticolonialism as spaces of cosmopolitan encounter which were discussed in the last chapter. Whilst not as dramatic (in some ways) as the encounter between Aurobindo and Alfassa in the last chapter, the movements of Acharya indicate something of the deep social ties that helped to bind together the international anticolonial 'movement' as it occurred in and across the spaces which it was shaping in the first half of the twentieth century.

Conclusions

This chapter is the most territorially 'distant' from the Pondicherry 'Gang' in many ways, but it also shows how tiny French Pondicherry was intimately connected to much wider and incredibly dynamic movements which existed globally. On one level, reading M.P.T. Acharya's life tells us an extensive spatial story of how far the 'reach' of the members of the Pondicherry 'Gang' could be. Acharya was only in Pondicherry for a short time, but the networks he inhabited in his political life moved him from provincial South India into a diverse world of international revolutionary currents. It is tempting to read him then as a precursor to the networked geographies of 'globalised' social movements. However, we should sound a note of caution here. In their introduction to their book on the internationalist nature of the interwar years, Raza, Roy and Zachariah's (2015, p. xxii) call for care when using these terms, as they rightly state that:

> old clichés of nationalist or communist histories are in danger of giving way to new clichés of the transnational, the global, or the cosmopolitan

In this chapter, whilst addressing an avowedly international life story, I have argued that M.P.T. Acharya's anticolonial struggle was not simply a cliché of mobility or cosmopolitan encounter. Instead, examining someone like Acharya shows how deeply fragmented and precarious living within these 'international' spaces of revolutionary activity were. We can see that, at stages and times of his life, moments of internationalism such as when trying to organise a sympathetic solidarity from Constantinople in 1911, or a space of loving encounter across difference in his marriage to Nachman. However, we also see moments of failure, such as the disastrous move to support the Rif rebellion, but also the precarity of his life – his choice to move away from the USSR due to his anarchist beliefs, unlike his committed friend Chatto, meant that he survived, albeit in straitened circumstances.

This again shows the different ways in which anticolonialism was envisioned and shows how internationalist anticolonialism circulated in a more diverse range of spaces than it is often given credit for. This gives a sense by which internationalism was created across a variety of spatial practices and scales which are more complex than anticolonialism simply 'touching down' in a place. As I discussed in Chapter Two, and has been noted throughout this book, nationalism, whilst a dominant framing for anticolonial activities, was not the only one, and operated alongside international and cosmopolitan renderings of the world (Chatterjee 2016). Acharya's explicit rejection of these categories to embrace anarchism is unusual amongst the Indian independence movement, especially given the limited room for alternative ideologies at this time. Using some of the spatial understandings of social movements, this chapter has argued that we should examine the international collections of revolutionaries using some of the situated and nuanced geographies of internationalism which are attentive to how 'internationalism' is always routed through the places and spaces by which it is encountered (Hodder 2015). The very real encounters with Soviet or Bolshevik forms of communism ensured that people like Acharya, whose more open and egalitarian politics rejected Hindu-centric visions of India as well, was unable to countenance. What is clear here is how the anticolonial spaces which Acharya moved in were distinct and were eminently more than spaces of *ressentiment*, and this emphasises the core message of this book – that the geographies of anticolonialism were dynamic and diverse spaces through which those involved sought to make their visions of a better future a reality.

References

Abraham, I. (2015). "Germany has become Mohammedan": insurgency, long-distance travel, and the Singapore Mutiny, 1915. *Globalizations* 12 (6): 913–927. https://doi.org/10.1080/14747731.2015.1100850.

Acharya, M. (1991). *Reminiscences of an Indian Revolutionary* (ed. B. Yadav). New Delhi: Anmol.

Acharya, M. (2019). *We Are Anarchists: Essays on Anarchism, Pacifism, and the Indian Independence Movement, 1923–1953* (ed. O.B. Laursen). Edinburgh: AK Press.

Adi, H. (2013). *Pan-Africanism and Communism: The Communist International, Africa and the Diaspora 1919–1939*. Trenton, NJ: Africa World Press.

Ahuja, R. (2010). The corrosiveness of comparison: reverberations of Indian wartime experiences in German Prison Camps (1915–1919). In: *The World in World Wars* (ed. H. Liebau), 131–166. Leiden: Brill.

Almy, R.L. (2018). "More hateful because of its hypocrisy": Indians, Britain and Canadian Law in the Komagata Maru Incident of 1914. *The Journal of Imperial and Commonwealth History* 46 (2): 304–322. https://doi.org/10.1080/03086534.2018.1438964.

Anderson, B. (2007). *Under Three Flags: Anarchism and the Anti-Colonial Imagination*, 1e. London: Verso.

Bald, V. (2013). *Bengali Harlem and the Lost Histories of South Asian America*. Cambridge, MA: Harvard University Press.

Barooah, N.K. (2004). *Chatto*. Oxford: Oxford University Press.

Bayly, C.A. (2004). *The Birth of the Modern World*. Oxford: Blackwell.

Bayly, C.A. (2012). *Recovering Liberties: Indian Though in the Age of Liberalism and Empire*. Cambridge: Cambridge University Press.

Bernstein, L. (2017). Indian nationalists' cooperation with Soviet Russia in Central Asia: the case of M.P.T. Acharya. In: *Personal Narratives, Peripheral Theatres: Essays of the Great War (1914–1918)* (eds. A. Barker, M.E. Pereira, M.T. Cortez, et al.), 201–214. Cham: Springer.

Bernstein, L. (2018). The Great Little Lady of the Bombay Art World. In: *Transcending the Borders of Countries, Languages, and Disciplines in Russian Émigré Culture* (eds. C. Flamm, R. Marti and A. Raev), 143–158. Newcastle Upon Tyne: Cambridge Scholars Publishing.

Boehmer, E. (2015). *Indian Arrivals 1870–1915: Networks of British Empire*. Oxford: Oxford University Press.

Bose, A.C. (2002). *Indian Revolutionaries Abroad:1905–1927: Select Documents*. New Delhi: Northern Book Centre.

Bose, S. and Manjapra, K. (eds.) (2010). *Cosmopolitan Thought Zones: South Asia and the Global Circulation of Ideas*. Basingstoke: Palgrave Macmillan.

Brah, A. (1996). *Cartographies of Diaspora: Contesting Identities*. London: Routledge.

Chandra, B. et al. (1989). *India's Struggle for Independence*. New Delhi: Penguin.

Chatterjee, P. (2016). Nationalism, internationalism, and cosmopolitanism: some observations from modern Indian history. *Comparative Studies of South Asia, Africa and the Middle East* 36 (2): 320–334. https://doi.org/10.1215/1089201X-3603392.

Collins, M. (2012). Rabindranath Tagore and the politics of friendship. *South Asia: Journal of South Asian Studies* 35 (1): 118–142. https://doi.org/10.1080/00856401.2011.648908.

Davies, A.D. (2012). Assemblage and social movements: Tibet Support Groups and the spatialities of political organisation. *Transactions of the Institute of British Geographers* 37 (2): 273–286. https://doi.org/10.1111/j.1475-5661.2011.00462.x.

Davies, A.D. (2013). Identity and the assemblages of protest: the spatial politics of the Royal Indian Navy Mutiny, 1946. *Geoforum* 48: 24–32. https://doi.org/10.1016/J.GEOFORUM.2013.03.013.

Davies, A.D. (2014). Learning "large ideas" overseas: discipline, (im)mobility and political lives in the Royal Indian Navy Mutiny. *Mobilities* 9 (3): 384–400. https://doi.org/10.1080/17450101.2014.946769.

Derrick, J. (2008). *Africa's Agitators: Militant Anti-Colonialism in Africa and the West, 1918–1939*. London: Hurst & Co.

Elam, J.D. and Moffat, C. (2016). On the form, politics and effects of writing revolution. *South Asia: Journal of South Asian Studies* 39 (3): 513–524. https://doi.org/10.1080/00856401.2016.1199293.

Featherstone, D.J. (2005). Towards the relational construction of militant particularisms: or, why the geographies of past struggles matter for resistance to neoliberal globalisation. *Antipode* 37 (2): 250–271.

Featherstone, D.J. (2012). *Solidarity: Hidden Histories and Geographies of Internationalism*. London: Zed Books.

Featherstone, D. (2017). Anti-colonialism and the contested spaces of communist internationalism. *Socialist History* 52: 48–58.

Fischer-Tine, H. (2007). Indian Nationalism and the "world forces": transnational and diasporic dimensions of the Indian freedom movement on the eve of the First World War. *Journal of Global History* 2 (3): 325–344.

Gandhi, L. (2006). *Affective Communities: Anticolonial Thought, fin-de-siècle Radicalism, and the Politics of Friendship, Politics, History, and Culture* (eds. J. Adams and G. Steinmetz). Durham, NC: Duke University Press.

Ghose, A. (2002). *Bande Mataram: Political Writings and Speeches 1890–1908. Complete Works of Sri Aurobindo*, vol. 6 and 7. Pondicherry: Sri Aurobindo Ashram.

Griffin, P. (2018). Diverse political identities within a working class presence: revisiting Red Clydeside. *Political Geography* 65: 123–133. https://doi.org/10.1016/J.POLGEO.2018.06.002.

Halvorsen, S. (2015). Encountering occupy London: boundary making and the territoriality of urban activism. *Environment and Planning D: Society and Space* 33 (2): 314–330. https://doi.org/10.1068/d14041p.

Harvey, D. (1995). Militant particularism and global ambition: the conceptual politics of place, space, and environment in the work of Raymond Williams. *Social Text* 42: 69–98. https://doi.org/10.2307/466665.

Heehs, P. (1992). The Maniktala secret society: an early Bengali terrorist group. *The Indian Economic and Social History Review* 29 (3): 349–370.

Heehs, P. (2008). *The Lives of Sri Aurobindo*. New York: Columbia University Press.

Hodder, J. (2015). Conferencing the international at the World Pacifist Meeting, 1949. *Political Geography* 49: 40–50. https://doi.org/10.1016/j.polgeo.2015.03.002.

Hodder, J., Legg, S., and Heffernan, M. (2015). Introduction: historical geographies of internationalism, 1900–1950. *Political Geography* 49: 1–6. https://doi.org/10.1016/j.polgeo.2015.09.005.

Høgsbjerg, C. (2011). Mariner, renegade and castaway: Chris Braithwaite, seamen's organiser and Pan-Africanist. *Race & Class* 53 (2): 36–57. https://doi.org/10.1177/0306396811414114.

Hyslop, J. (2011). An "eventful" history of Hind Swaraj: Gandhi between the Battle of Tsushima and the Union of South Africa. *Public Culture* 23 (2): 299–319. https://doi.org/10.1215/08992363-1162048.

Jessop, B., Brenner, N., and Jones, M. (2008). Theorizing sociospatial relations. *Environment and Planning D: Society and Space* 26 (3): 389–401. https://doi.org/10.1068/d9107.

Jones, M. (2016). Polymorphic political geographies. *Territory, Politics, Governance* 4 (1): 1–7. https://doi.org/10.1080/21622671.2015.1125650.

Kapila, S. (2007). Self, Spencer and swaraj: nationalist thought and critiques of liberalism, 1890–1920. *Modern Intellectual History* 4 (1): 109. https://doi.org/10.1017/S1479244306001077.

Koopman, S. (2015). Social movements. In: *The Wiley Blackwell Companion to Political Geography* (eds. J. Agnew et al.), 229–351. Oxford: Wiley-Blackwell.

Laursen, O.B. (2018). Anarchist anti-imperialism: Guy Aldred and the Indian Revolutionary Movement, 1909–14. *The Journal of Imperial and Commonwealth History* 46 (2): 286–303. https://doi.org/10.1080/03086534.2018.1431435.

Laursen, O.B. (2019). MPT Acharya: a revolutionary, an agitator, a writer. In: *We Are Anarchists: Essays on Anarchism, Pacifism, and the Indian Independence Movement, 1923–1953*. Edinburgh: AK Press.

Laursen, O.B. (n.d.). "Anarchism, pure and simple": M. P. T. Acharya, anti-colonialism and the international anarchist movement. *Postcolonial Studies* Forthcoming.

Legg, S. (2014). An international anomaly? Sovereignty, the League of Nations and India's princely geographies. *Journal of Historical Geography* 43: 96–110. http://dx.doi.org/10.1016/j.jhg.2013.03.002.

Leitner, H., Sheppard, E., and Sziarto, K.M. (2008). The spatialities of contentious politics. *Transactions of the Institute of British Geographers* 33 (2): 157–172. https://doi.org/10.1111/j.1475-5661.2008.00293.x.

Maclean, K. (2015). *A Revolutionary History of Interwar India: Violence, Image, Voice and Text.* London: Hurst & Company.

Maclean, K. and Elam, J.D. (2013). Reading revolutionaries: texts, acts, and afterlives of political action in late colonial South Asia. *Postcolonial Studies* 16 (2): 113–123. https://doi.org/10.1080/13688790.2013.823259.

McAdam, D., Tarrow, S., and Tilly, C. (2001). *Dynamics of Contention.* Cambridge: Cambridge University Press.

McFarlane, C. (2009). Translocal assemblages: space, power and social movements. *Geoforum* 40 (4): 561–567.

McFarlane, C. (2011). *Learning the City: Knowledge and Translocal Assemblage.* Oxford: Wiley-Blackwell.

Miller, B. (2000). *Geography and Social Movements: Comparing Antinuclear Activism in the Boston Area.* Minneapolis: University of Minnesota Press.

Mishra, P. (2013). *From the Ruins of Empire: The Revolt Against the West and the Remaking of Asia.* London: Penguin.

Mukherjee, S. (2018). *Indian Suffragettes: Female Identities and Transnational Networks.* Oxford: Oxford University Press.

Nicholls, W.J. (2007). The geographies of social movements. *Geography Compass* 1 (3): 607–622. https://doi.org/10.1111/j.1749-8198.2007.00014.x.

Nicholls, W.J. (2008). The urban question revisited: the importance of cities for social movements. *International Journal of Urban and Regional Research* 32 (4): 841–859. https://doi.org/10.1111/j.1468-2427.2008.00820.x.

Nicholls, W., Miller, B., and Beaumont, J. (eds.) (2013). *Spaces of Contention: Spatialities of Social Movements.* Farnham: Ashgate.

Petersson, F. (2014). Hub of the anti-imperialist movement. *Interventions* 16 (1): 49–71. https://doi.org/10.1080/1369801X.2013.776222.

Prashad, V. (2008). *The Darker Nations: A People's History of the Third World.* New York: The New Press.

Ramnath, M. (2011a). *Decolonizing Anarchism: An Antiauthoritarian History of India's Liberation Struggle, Anarchist Interventions.* Edited by I. for A. Studies. Edinburgh: AK Press.

Ramnath, M. (2011b). *Haj to Utopia: How the Ghadar Movement Charted Global Radicalism and Attempted to Overthrow the British Empire.* Berkeley: University of California Press.

Raza, A., Roy, F., and Zacharia, B. (2015). Introduction. In: *The Internationalist Moment* (eds. A. Raza, F. Roy and B. Zachariah), xi–xli. New Delhi: SAGE Publications India.

Routledge, P. (2017). *Space Invaders: Radical Geographies of Protest.* London: Pluto.

Roy, A.G. and Sahoo, A.K. (2016). The journey of the Komagata Maru: national, transnational, diasporic. *South Asian Diaspora* 8 (2): 85–97. https://doi.org/10.1080/19438192.2016.1221201.

Sharp, J.P. et al. (eds.) (2000). *Entanglements of Power: Geographies of Domination/Resistance.* London: Routledge.

Subramanyan, C.S. (1995). *MPT Acharya: His Life and Times: Revolutionary Trends in the Early Anti-Imperialist Movements in South India and Abroad.* Madras: Institute of South Indian Studies.

Tarrow, S. (1994). *Power in Movement: Social Movements, Collective Action and Politics.* Cambridge: Cambridge University Press.

Tickell, A. (2012). Scholarship terrorists: the India House Hotel and the "Student Problem" in Edwardian London. In: *South Asian Resistances in Britain 1858–1947* (eds. S. Mukherjee and R. Ahmed), 3–18. London: Continuum.

Tilly, C. and Wood, L.J. (2012). *Social Movements: 1768–2012*, 3e. London: Routledge.

Venkatachalapathy, A. R. (2018) 'AR Venkatachalapathy on Subramania Bharati vs Rabindranath Tagore', *Hindustan Times*, 12 May.

Visram, R. (2002). *Asians in Britain: 400 Years of History*. London: Pluto.

White, S. (1976). Colonial revolution and the Communist International, 1919–1924. *Science & Society* 40 (2): 173–193.

Worley, M. (ed.) (2004). *In Search of Revolution: International Communist Parties in the Third Period*. London: I.B. Taurus.

Yadav, B.D. (1992). *Madame Cama: A True Nationalist*. New Delhi: Anmol.

Zacharia, B. (2015). Internationalisms in the interwar years: the travelling of ideas. In: *The Internationalist Moment*, 1–21. New Delhi: SAGE Publications India.

Chapter Eight
Conclusion: The Necessity of a Geographical Anticolonial Thought, or Why Anticolonialism Still Matters

The Pondicherry 'Gang' was, in the grand scheme of anticolonial resistance, a minor example compared to the huge mobilisations that followed later in India and elsewhere in the twentieth century. However, to think of it in terms of scale is, as I hope is clear by now, to miss the point. Each of the four individuals who formed the focal point of each chapter of this book, Subramania Bharati, V.O.C. Pillai, Aurobindo Ghose and M.P.T. Acharya, could also be counted as failures in purely *Political* terms – none of them decisively 'won' their fight against colonialism, with the possible exception of Aurobindo, whose decision to focus on the evolution of the human spirit defied what the colonial authorities (and often he and his followers) understood as *Politics*. Bharati and Pillai both did not live to see independence, and even though Acharya did, all of this latter three lived out their lives in straitened, often poverty-stricken, circumstances, victims of a variety of the various structural violences which colonialism imposed upon those who attempted to resist it. However, in the longer term, all, with the possible exception of Acharya, have become recognised as important figures in the struggle against British colonialism in South Asia, although, again, for Acharya, this is changing (Acharya 2019).

However, the point is not whether the spaces of anticolonialism which were produced by these men who passed through Pondicherry were immediate Political 'successes', but rather how they cleared space for alternative political imaginaries to be mobilised. It is better instead to follow Manu Goswami's (2012) recognition that the mode(s) of anticolonial revolutionary thought were and are inherently about thinking in often utopian terms about the possible

Geographies of Anticolonialism: Political Networks Across and Beyond South India, c. 1900–1930, First Edition. Andrew Davies.
© 2020 Royal Geographical Society (with the Institute of British Geographers). Published 2020 by John Wiley & Sons Ltd.

future to come. The geographies which these imaginary futures shaped were as diverse as the possibilities which lay beyond the existing world of colonialism and imperialism which these men inhabited, and which they decided to challenge at great personal cost.

V.O.C. Pillai's establishment of the Swadeshi Steam Navigation Company (SSNCo) in 1906 should attract more attention simply because of its importance in starting some of the largest mobilisations against colonialism in South India. However, as I argued in Chapter Five, the SSNCo can be read archipelagically to destabilise the 'landed' nature of anticolonialism. This is more than recognising that the sea and its associated ship-spaces provided a vital conduit for the trans-mission of anticolonial materials around the world. It is about understanding how the intersection between land and sea was a space that was productive of antico-lonial activity, even in areas which were supposedly 'backwaters' of Political agi-tation. Creating and developing a national economy was, in itself, a standard desire of the *swadeshi* rulebook for its vision of the future, and so in that regard the SSNCo was following a set repertoire of contention. But by moving things to sea, it played a game of regional/transnational connections that meant it was most definitely a threat to colonial vested interests, despite its small size. Here, we could think of the SSNCo as something akin to the Zapatista's present day calls for 'the war of the flea' – using deliberately tiny enterprises or activities to both push back against colonial authority and expose the unequal nature of the struggle which anticolonialists are engaged in.

Subramania Bharati's distinctly Tamil approach to creating a vision of India after colonialism was polyglot and innovative. His eccentric behaviour masked a visionary ability to appropriate existing texts and media to suit his needs – importing the use of cartoons for political and satirical commentary into India, or reinventing the prose and poetics of Tamil writing to be sharp and pointed enough to work in the twentieth century. This Tamilian context is not only the most defi-nitely rooted in the Tamil cultural-milieu of the fin-de-siècle but also indicative of wider changes affecting Indian anticolonialism. This is a placed anticolonialism, but most definitely not one which is parochially local. Bharati's life shows how seeming 'backwaters' like Pondicherry were intimately connected to global circuits of revolutionary and anticolonial activity.

The next chapter explored some of the ways in which the colonial/modern definition of the political can be contested through an engagement with the deco-lonial spiritual. Here it is important to re-state that this book's emphasis on the anticolonial does not mean that it is in competition with the postcolonial and decolonial – rather, they are complimentary and can be used in mutually benefi-cial ways. Whilst Aurobindo Ghose would probably not have thought about his dedication towards spiritual reinvention in these terms, when his move from Chandernagore to Pondicherry is read decolonially, it softens and makes porous the supposed divide between his 'anticolonial' and his 'spiritual' lives (as Heehs might call them). This provides space to understand Aurobindo's life with more tolerance than relying on the spiritual/secular debate, as it means that we treat Aurobindo's spirituality, including that which is unexplainable to many, with as much thought as we do his political writings.

The final chapter again focussed on someone who had distinct phases to their revolutionary life. M.P.T. Acharya's life at its most basic level shows the astonishing spatial mobility that was open to the revolutionary anticolonialists of this era. However, it also showed the diverse patterns of overlapping ideologies which occurred during this movement as well. That anarchism proved to be how Acharya defined his politics should not be especially surprising, but it does speak to how minor Indian anarchisms were, or, more correctly, how little attention has been directed at them. Part of the challenge here, as Ole Laursen has pointed out, is that highly mobile individuals like Acharya left their 'archives' scattered around the globe, often in different languages. This situation is changing, and the renewed interest in people like Acharya means that the currents of Indian anticolonialisms which were not only nationalist in thought and practice are gradually becoming more visible.

These four men and their various relationships have exposed certain aspects of how anticolonialism functioned in the spaces associated with Pondicherry. It is impossible to summarise the full shape and scope of the revolutionary ideologies and practices which were connected to Pondicherry, and nor should there be an attempt to write the 'complete' or 'true' story. The radical changes which occurred within the lives of the men and women who can be associated with the Pondicherry 'Gang' mean that each was relationally constituted and altered their positions and behaviours accordingly. It is thus important to treat these individuals, and how we understand them given the partial and fragmented archives we have of them, as difficult and contradictory characters. Here I also want to note again that this is why the terminology I have used to describe the 'Gang' has varied. I would categorise them all as 'anticolonial'; however, they viewed themselves as nationalists, revolutionaries, anarchists, gurus and more. Thus, I have tried to remain somewhat true to their positions in the writing of this book but have tended to emphasise the anticolonial given the purpose of my argument.

A related point to the partial and fragmentary pictures of these individuals is to recognise the problematic nature of some of the positions held by the people within this book. Whilst I agree with Heehs (2006) that figures like Aurobindo have been appropriated by both left and right in India and beyond, it is important to recognise that the revolutionary anticolonialists in this book were associated with people like Bal Gangadhar Tilak and V.D. Savarkar whose Hindu-centric views and contribution to religious revivalism have contributed to today's often toxic Hindu-nationalist politics in India. However, despite this, it is important to return to the progressive imperative that lies at the heart of anticolonialism that was outlined by Leela Gandhi (Gandhi 2007, 2011). Anticolonialism should always be aligned towards the politics of emancipation, and this is, I think, true of the majority of the people in this book.

This also leads to thinking about how academics, their books and their contents form part of the wider culture wars related to colonialism which are currently taking place. The reaction in India to Heehs' and Doniger's books about Aurobindo and Hinduism, respectively, was noted in Chapter Six. In United Kingdom and India, the publication of Shashi Tharoor's polemic about the negative conse-quences of British rule in India has provoked debate about historical accuracy (Tharoor 2017; Allen 2018; Roy 2018). As I was completing the revisions on this book, the centenary of the Amritsar Massacre was marked in the United Kingdom by a Channel 4 TV documentary by the novelist Sathnam Sanghera which again exposed the inability of sections of the UK's public opinion to adequately recog-nise the violent history that comes along with colonial rule. Part of this is no doubt fuelled by an increased culture of seeking Manichean answers to complex prob-lems, driven especially by a need to condense ideas into easily digestible, social media-friendly, packaging, but is also driven by the spread of 'fake news' and other negative and socially regressive practices. Discussions of how both colonialism and anticolonialism took place are still vitally important and are not just limited to the United Kingdom–India context either, as the variety of decolonial move-ments erupting around the world shows. Treating the men and women who passed through Pondicherry as imperfect and nuanced human beings is not simply a matter for the historical records, but it is rather an important reminder about why the *p*olitics of past and present anticolonialisms matter.

The geographies of anticolonialism then remain important and necessary to our understandings of the world around us. Colonialism, and its close relation imperialism, continues to create toxic relationships between us as human beings, dehumanising migrants, promoting ideologies of racial superiority, and generally poisoning the ability for humans to live to their full potential in a world free of hatred. Understanding past and present anticolonialisms helps to expose the rottenness at the heart of imperialist gestures. As I argued in the Chapters One, Two and Three of this book, thinking geographically about anticolonialism is important yet is often forgotten in our rush to engage with the latest disciplinary buzz-word as we are driven towards ever faster rates of publication. This approach is always about more than saying that there were a 'variety' of anticolonialisms present within the world which were spatially contingent. Anticolonialism is, to me, a particular way of engaging with the world which requires a different approach to other ways of engaging with colonialism. This should involve, as many geogra-phers and others are doing, re-examining how colonialism plays a role in setting out our disciplinary boundaries and concepts in a way which limits or excludes those who do not necessarily conform to them. Examining past colonialisms, as this book has, allows us to apply some hindsight to these activities and to assess their 'failures/successes', as well as to map out their various consequences and the spatialities that were generated as a result of them.

But, more than just thinking about the past, thinking 'anticolonially' shifts political spaces to the forefront of thinking about resistance to colonialism and imperialism – this is something that the more culturally/representationally focussed postcolonial and epistemologically focussed or present-centric decolonial have done in part, but returning to and reinvigorating anticolonial geographies offers something that we are in danger of, if not forgetting, at least underplaying – the importance of challenging the political frameworks which were established by colonialism but which we still continue to use. This is more than combatting the postcolonial 'nation-state' as the most visible reminder of the legacies of colonialism, but rather it is continuing to engage with and hold to account the centres of power established by colonialism, as well as the conceptions of the world that continue to reproduce colonial practices. Used alongside and in engagement with post- and decolonial approaches to space, we can continue to interrogate the various forms which resistance to colonialism and imperialism can take. As recent debates within disciplinary geography but also in society more broadly, not only in 'the West', have (sadly) shown these struggles are very much alive, and reckoning with our colonial and imperial pasts is by no means finished.

References

Acharya, M. (2019). *We Are Anarchists: Essays on Anarchism, Pacifism, and the Indian Independence Movement, 1923–1953* (ed. O.B. Laursen). Edinburgh: AK Press.

Allen, C. (2018). Who owns India's history? A critique of Shashi Tharoor's inglorious empire. *Asian Affairs* 49 (3): 355–369. https://doi.org/10.1080/03068374.2018.1487685.

Gandhi, L. (2007). Postcolonial theory and the crisis of European man. *Postcolonial Studies* 10 (1): 93–110. https://doi.org/10.1080/13688790601153180.

Gandhi, L. (2011). The Pauper's gift: postcolonial theory and the new democratic dispensation. *Public Culture* 23 (1): 27–38. https://doi.org/10.1215/08992363-2010-013.

Goswami, M. (2012). Imaginary futures and colonial internationalisms. *The American Historical Review* 117 (5): 1461–1485. https://doi.org/10.1093/ahr/117.5.1461.

Heehs, P. (2006). The uses of Sri Aurobindo: mascot, whipping-boy, or what? *Postcolonial Studies* 9 (2): 151–164. https://doi.org/10.1080/13688790600657827.

Roy, T. (2018). Inglorious empire: what the British did to India. *Cambridge Review of International Affairs* 31 (1): 134–138. https://doi.org/10.1080/09557571.2018.1439321.

Tharoor, S. (2017). *Inglorious Empire: What the British did to India*. London: Hurst & Co.

Bibliography

Cambridge South Asia Archive

Madras Native Newspaper Reports 1907–1908.
Ashe Papers.

National Archives of India, New Delhi

Private Archives

'Diary of His Britannic Majesty's Consul at Pondicherry regarding Shri Aurobindo's Movements and visitors ay Pondicherry, July 1910 – 3ʳᵈ November 1914), History of the Freedom Movement Collection (Serial No. 44) Vol. 2 (1885–1919) Section B, File No ½.

Public Records

Home Political, Branch A April 1908, Nos. 79–87 'Disturbances in the Tinnevelly and Tuticorin Districts'.
Home Political, Branch A June 1908, No. 95 'History of the Disturbances in Tinnevelly and Tuticorin in March 1908'.

Geographies of Anticolonialism: Political Networks Across and Beyond South India, c. 1900–1930, First Edition. Andrew Davies.
© 2020 Royal Geographical Society (with the Institute of British Geographers). Published 2020 by John Wiley & Sons Ltd.

Home Political, Branch B July 1908, No. 40 'Measures taken to Prevent the Introduction of the Gaelic American and Indian Sociologist into Chandernagore. Delegation of Authority to Certain Additional Officers of the Post Office of India to Search or Cause to be made for Copies of the Gaelic American and the Indian Sociologist'.

Home Political, Branch A, December 1908, Nos. 6–14 'Prosecution of Swadesamitran and India Newspapers under Sections 124-1, 153-A and 505 of the India Penal Code'.

Home Political (Confidential) Branch A, August 1919, No. 44–52 'Notice on Pondicherry and Chandernagore as Centres of Seditious Agitation'.

Home Political, Branch B, November 1909, Nos. 32–41 'Weekly Reports of the Director, Criminal Intelligence and the Government of the Punjab on the Political Situation, during the Month of October 1909'.

Home, Political, Branch B, December 1909, No. 37. 'Movements of M.P. Tirumalaichari, formerly proprietor, publisher and Editor of a Tamil newspaper called India, and Sukkagar Dutt'.

Home, Political Branch B, Feb. 1910, Nos. 143–145 'Proposed prohibition under the Indian Press Act (I of 1910) of the importation into India of the *India* and *Suryodayam* newspapers of Pondicherry'.

Home Political (Deposit) April 1910, No. 20 'Importation of arms into Chandernagore and Pondicherry, French Arms Act'.

Home Department, Political, Branch B, May 1910, Nos. 191–193 – 'Selections from the weekly native newspapers in the Madras Presidency for the week ending 2nd April 1910'.

Home, Political (Deposit), March 1911, No. 12 'Dissemination of Seditious Literature through Pondicherry and Chandernagore'.

Home, Political (Deposit), April 1911, No. 7 'Information regarding the activities of the Revolutionary Party in London and Paris'.

Home, Political, Branch B, July 1911, Nos. 41–42 'Letter from His Excellency M. Martineau, Governor of the French settlements in India, condemning the murder of Mr. Ashe.

Home Political, Branch A, October 1911, Nos. 114–117 'Issue of Orders under Section 26(1) of the Post Office Act for the retention by the Madras Criminal Investigation Department of certain postal articles addressed to certain revolutionaries'.

Home, Political, Branch A, May 1912, Nos. 28–29 'Interchange of information between the French and British Officers in connection with the activities of the Anarchists at Chandernagore'.

Home, Political, Branch A, June 1912, Nos. 41–68 'Murder of Mr Ashe, Collector of Tinnevelly. Judgements in the Ashe Murder and Conspiracy cases'.

Home Political, Branch N, August 1913, No. 61 'History Sheet of Madame Cama'.

Home Political, Branch A, December 1913, Nos. 15–16 'Proposed Cession of Chandernagore to the British Crown'.

Home, Political A, May 1918, Nos. 308–310 'French Settlements in India – Negotiations with the French Government in regard to the expulsion of seditious Indians from Chandernagore'.

Sri Aurobindo Ashram Archives, Pondicherry

Copy of Cartoon 'The Rahu which came along to swallow the Aurobindo Sun is slinking away' *India* (Intiya) May 1909.

Tamil Nadu Archives, Chennai

GO No. 1407–1408 10/08/1907 Judicial (Confidential) 'Political Agitations – police should openly attend all political meetings etc.'.

GO No. 1103, 11/08/1908, Judicial (Confidential) 'Prosecution of Sri M. Srinivasa Iyengar Editor of the India for having published seditious articles'.

GO No. 1729, 29/12/1908, Judicial (Confidential) 'Powers of the Police in regard to regulating and prohibiting assemblies and processions'.

GO No. 44, 12/01/1909, Judicial (Confidential) 'Prosecutions for Sedition in Madras Presidency'.

GO No. 408 Judicial Department (Mis. Confidential) Dated 13/03/12 'Informing the Inspector-General of Police that the thanks of Government have been conveyed to the Governor of French Settlements for the assistance rendered by the Commandant of the Police.

GOs 1010–1014 History of the freedom Movement Papers.

GO 1014 Judicial (Confidential) Dated 24/06/1912 'Recording correspondence regarding the retention of British Police in Pondicherry and the conduct of Inspector Siva Chidambaram Pillai'.

GO No. 1335 Judicial Department (Confidential) Dated 21/08/1912 'Recording correspondence regarding the watching of political suspects in Pondicherry and requesting the Inspector-General of Police to report again on 01/11/1912 whether any reduction in the Pondicherry force can be affected'.

Index

Acharya, M.P.T.
anarchist position of, 153–154
attempt to support Rif in North Africa, 150–151
Constantinople, letter to VVS Aiyer from, 137–138
fragmented archival record of, 139
marriage to Magda Nachman, 139, 153
movements through Europe, Asia and America of, 151–154
organisation of protests in Madras, 1908, 94
publishing work with Subramania Bharati in Madras and Pondicherry, 76, 100, 103
revolutionary work in Berlin, 147
work and life after return to India, 153–154
Acharya, Srinivas, 60, 103, 105, 108, 110
Aiyar, V.V.S.
activities in London, 145
activities in Pondicherry, 105, 109
involvement in revolver training in Pondicherry, 108
letter from MPT Acharya in Constantinople to, 137–138
Albert, Joseph 'Libertad', 138
Aldred, Guy, 153

Alfassa, Mirra
Arrival of in Pondicherry, 126
background in mysticism, 126
building of Auroville, 127–128
establishment of Sri Aurobindo Ashram, 127
relationship with Aurobindo, 126–127
Alipore bomb trial, 125
Ambedkar, B.R, 29, 127
Anarchism
links to anticolonial thought, 32–33
and M.P.T. Acharya, see Acharya, M.P.T
Anderson, Benedict, 32, 138
Anger, Roger, 127
Anticolonialism
difference to postcolonialism and decoloniality
and politics in Urban spaces, 96
and relation to 'successful' political activities, 161, 164
as more than *ressentiment*, 20–21, 23, 134, 154
lack of easy definition, 24–25
unruly ethics of, 33, 163
Apter, Emily, and unexceptional politics, 24

Geographies of Anticolonialism: Political Networks Across and Beyond South India, c. 1900–1930, First Edition. Andrew Davies.
© 2020 Royal Geographical Society (with the Institute of British Geographers). Published 2020 by John Wiley & Sons Ltd.

Criminal Investigation Department,
Government of Madras, 77, 103,
131, 132
Curzon, George, Viceroy of India
(1899–1905), 73

Dacoity (violent armed robbery), 124
Das, Hem Chandra, 124, 138
Deep-relation of solidarity between
colonised peoples, 24, 33, 125
Department of Criminal Intelligence,
Government of India, 151
Dhingra, Madan Lal, 28, 145, 150
Doniger, Wendy, 130, 164
Drain Theory, 40, 73, 75
Dupleix, SS (Steamship), 103,
106, 125

Extremist nationalist members of Indian
National Congress, 25, 44, 92, 100,
122, 124, 147

Fanon, Frantz, 8, 21, 22, 29
Featherstone, David, 52–55, 121, 142

Gallia, S.S., 77, 83
Gandhi, Leela, 27, 33–36, 120, 128–129,
163
Gandhi, Mohandas (Mahatma)
disagreement with extremists, 25, 28
and *Hind Swaraj*, 25–31
on non-violence, 25–28, 41–42
Geographies of religion, 119
Gender, role in interwar politics, 48
Ghadar movement, 146–148
Ghose, Aurobindo
and *Adesh's*, 115, 117, 125
appropriation of ideas in postcolonial
India, 121, 163
arrival and life in Pondicherry, 125
colonial police and misinterpretations
of writing of, 122–123
essentialist aspects of life and
thought, 131
increasing spirituality in Alipore
Jail, 125
and integral yoga, 14, 126, 133

rejection of spiritual/political
dichotomy, 130–133
on passive resistance, 28, 122–124
theoretical writing of, 122–123,
129–133
Ghose, Barin (brother of Auroindo), 122,
123, 124, 125, 138
Goldman, Emma, 138
Government of India (GoI), 29, 42, 50,
81, 82, 92, 94, 95, 96, 97, 103, 104,
107, 110, 111, 132

Har Dayal, Lala, 146
Hardiman, David, on non-violence
and subaltern studies, 53–54,
121–122
Heehs, Peter, controversy over *The Lives
of Sri Aurobindo*, 129–131, 162
Hind Swaraj, see Gandhi, Mohandas
Hindu revivalism, contested links to
anticolonial politics, 75, 101, 122,
129, 138
Hodder, Jake, 143

India (Tamil Publication), 76, 77, 78,
97, 100–101, 102, 103, 106,
125, 149
India House, London, 28, 29, 97,
145–146, 150, 151, 153
Indian National Congress, 1, 26, 42, 44,
48, 99, 100, 123
Indian National Congress, extremist wing
of (pre-1908), 25, 29, 43, 68, 91,
92–93, 100, 108, 121, 144, 147
Indian National Congress, moderate wing
of (pre-1908), 25, 44, 73, 92–93,
100, 122, 123
Indian Sociologist (publication), 29, 97,
105, 145, 146, 153
Intelligence agencies (criminal and
political)
in French India (Pondicherry), 104,
108, 132
and development of surveillance
techniques, 49, 76, 96
and interception of anticolonialists'
mail, 108, 138